BACKYARD
HOMESTEADING

CRE▲TIVE
HOMEOWNER®

BACKYARD
HOMESTEADING

A BACK-TO-BASICS
GUIDE TO *SELF-SUFFICIENCY*

DAVID TOHT

CREATIVE HOMEOWNER®, Upper Saddle River, New Jersey

BACKYARD HOMESTEADING

AUTHOR	David Toht
MANAGING EDITOR	Fran J. Donegan
PROOFREADER	Sara M. Markowitz
PHOTO COORDINATOR, DIGITAL IMAGING	Mary Dolan
INDEXER	Schroeder Indexing Services
DESIGN AND LAYOUT	David Geer
FRONT COVER ILLUSTRATION	Robin Moline
INTERIOR ILLUSTRATIONS	Rebecca Anderson, Ian Worpole

CREATIVE HOMEOWNER

VICE PRESIDENT AND PUBLISHER	Timothy O. Bakke
MANAGING EDITOR	Fran J. Donegan
ART DIRECTOR	David Geer
PRODUCTION COORDINATOR	Sara M. Markowitz

Current Printing (last digit)
10 9 8 7 6 5 4

Manufactured in the United States of America

Backyard Homesteading, First Edition
Library of Congress Control Number: 2010936952
ISBN-10: 1-58011-521-7
ISBN-13: 978-1-58011-521-6

CREATIVE HOMEOWNER®
A Division of Federal Marketing Corp.
24 Park Way
Upper Saddle River, NJ 07458
www.creativehomeowner.com

DAVID TOHT, once the proud owner of eight brown-egg biddies, has more than 60 how-to books to his credit. He considers harvesting a sun-warmed tomato from his own garden one of life's sweetest pleasures. He and his wife, Rebecca, live in Olympia, Washington.

Acknowledgments

Special thanks to those who generously shared their expertise, including Paul Basile, Chuck Bauer, Kristy Bell, Jim Berg, Stefanie Booth, Chana Sale Barn, Gurney Davis, Michelle Jackson, T.J. and Stephanie Johnson, Jeff and Kari Ann Holcomb, Phoebe Larson, Maggie Leman, Dori Lowell, Leslie Miller, Gita Moulton, National Presto Industries, Inc., Susan O'Neill, Steve and Kristine Orth, Megan Paska, Jack Robertson, Ernie Schmidt, Donna Semasko, Ben Shapiro, Joel Sommer, Susan Stillinger, Rob Thoms, Christine E. Wiegand

Safety First

Though all concepts and methods in this book have been reviewed for safety, it is not possible to overstate the importance of using the safest working methods possible. What follows are reminders—do's and don'ts for yard work and landscaping. They are not substitutes for your own common sense.

- *Always* use caution, care, and good judgment when following the procedures described in this book.

- *Always* determine locations of underground utility lines before you dig, and then avoid them by a safe distance. Buried lines may be for gas, electricity, communications, or water. Start research by contacting your local building officials. Also contact local utility companies; they will often send a representative free of charge to help you map their lines. In addition, there are private utility locator firms that may be listed in your Yellow Pages or online. Note: previous owners may have installed underground drainage, sprinkler, and lighting lines without mapping them.

- *Always* read and heed the manufacturer's instructions for using a tool, especially the warnings.

- *Always* ensure that the electrical setup is safe; be sure that no circuit is overloaded and that all power tools and electrical outlets are properly grounded and protected by a ground-fault circuit interrupter (GFCI). Do not use power tools in wet locations.

- *Always* wear eye protection when using chemicals, sawing wood, pruning trees and shrubs, using power tools, and striking metal onto metal or concrete.

- *Always* read labels on chemicals, solvents, and other products; provide ventilation; heed warnings.

- *Always* wear heavy rubber gloves rated for chemicals, not mere household rubber gloves, when handling toxins.

- *Always* wear appropriate gloves in situations in which your hands could be injured by rough surfaces, sharp edges, thorns, or poisonous plants.

- *Always* wear a disposable face mask or a special filtering respirator when creating sawdust or working with toxic gardening substances.

- *Always* keep your hands and other body parts away from the business ends of blades, cutters, and bits.

- *Always* obtain approval from local building officials before undertaking construction of permanent structures.

- *Never* work with power tools when you are tired or under the influence of alcohol or drugs.

- *Never* carry sharp or pointed tools, such as knives or saws, in your pockets. If you carry such tools, use special-purpose tool scabbards.

Contents

Introduction

Black earth turned into yellow crocus,
is undiluted hocus-pocus
—Piet Hein

My first brush with backyard homesteading came when we decided
to supplement our burgeoning vegetable garden with a flock of brown-egg
layers. We started with eight chicks, placing them in a cardboard box in the
basement with plenty of sawdust for them to scratch in and a lamp placed just so to
keep them warm. Our kids loved to hold the chicks; a great way for them to learn
how to be gentle with young animals while the chicks became comfortable with
humans. As the chicks grew into pullets, we moved them to a pen next to
the garage. I cut a doorway into the side of the garage and installed
some nesting boxes and a chicken-wire barrier.

Eventually, we gathered our first egg, a small one. We referred to it as our $50 egg because of all the feed we put into the flock. Soon our eight layers were in full swing, producing seven or eight eggs a day, enough to keep us more than supplied, plus a couple of dozen to sell at the office each week.

The hens were a bunch of sweeties, blowsy in their red-brown plumage. They always came out to greet us, and whenever we'd approach them with a handful of weeds from the garden, they would tumble over each other to get to the best pickings. We had quite a cycle going. Our part of the deal was to supply water, feed, fresh greens, and some crushed oyster shells. The chickens gave us eggs in return. Their manure went into the compost pile, which in turn fed the garden, starting the cycle all over again.

We kept them for a year, and it was a great run, but storm clouds loomed. Our hens were illicit biddies: a local ordinance forbade keeping any form of livestock. One day our alderman called and very nicely said someone had complained and we'd have to get rid of the flock. I argued that the hens made less noise and manure than a single German Shepherd, but of course, an ordinance is an ordinance. He gave us a few weeks to find them a new home. A farm family happily took them in.

Today, it is a rare bird that finds itself on the wrong side of the law. Motivated citizens' groups, enlightened municipal leaders, and a groundswell of enthusiasm for backyard farming have created exponential growth in the number of cities tolerant of homegrown livestock. And it is not just chickens. The gate has swung open wide to admit goats, sheep, and even pigs and cows. Bees have lost their stigma, and hives now stand proud on urban rooftops.

So you've chosen a great time to embark on the adventure of producing your own food, and this book is intended to help. In it you'll learn the basic techniques required and what types of crops and livestock might be right for your lot, skills, and climate. While *Backyard Homesteading* is far from being the last word on any of the topics covered—if you seriously get into chickens, goats, or bees, you'll fill shelves loaded with specialized books and magazines—it will get you started.

What can you expect on this adventure? As I spoke with scores of backyard homesteaders, consistent themes popped up. Most are pleased to achieve a degree of self-sufficiency in growing their produce, unhampered with doubts about what went into the food they put on their tables. They're glad to find ways to be kinder to mother earth, sustaining their food supply with minimal dependence on petroleum products, building up the soil as they take from it. Most love the honest toil involved in a farmstead, away from the flickering screen of a computer. Tilling the soil and turning compost strikes them as much more sensible exercise than a trip to the gym. They enjoy the friends they make along the way, especially the old hands whose passion for farming is so contagious. Lastly, they're fascinated by animals and love the adventure of partnering with other creatures.

And of course, they love engaging with the mystery that is life itself. Even a simple 4 × 8-foot garden plot can be loaded with discovery—the wonder of tiny pea sprouts emerging from the soil, the momentous presence of a gigantic zucchini lurking under a plant's canopy. (See pages 12–13 for ideas for homestead plans.) Animals provide even more wonder. My friend Ernie Schmidt, a veteran backyard farmer, recalls listening to a hive of happy bees on summer evenings. "It is just uncanny," he says. "They sound like the wind in a forest canopy. There is something magical about that noise."

And maybe that is the greatest allure—being part of the magic. Life, rather than being packaged in plastic wrap, is there in all its rich variety, surprise, and generosity. Right in your own backyard.

Dave Toht

Urban "Homestead"

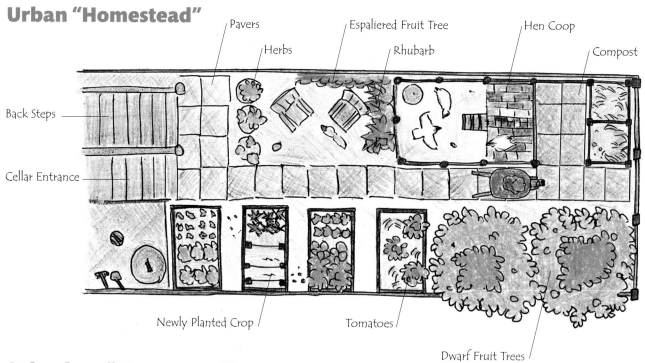

Pavers

Herbs

Espaliered Fruit Tree

Rhubarb

Hen Coop

Compost

Back Steps

Cellar Entrance

Newly Planted Crop

Tomatoes

Dwarf Fruit Trees

Suburban "Homestead"

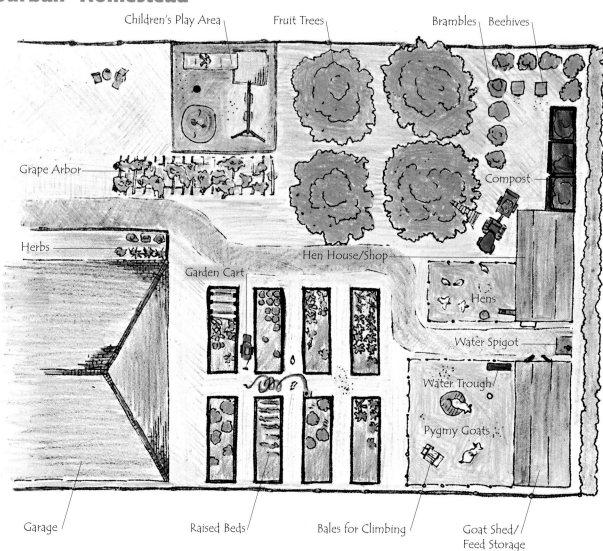

Children's Play Area

Fruit Trees

Brambles

Beehives

Grape Arbor

Compost

Herbs

Garden Cart

Hen House/Shop

Hens

Water Spigot

Water Trough

Pygmy Goats

Garage

Raised Beds

Bales for Climbing

Goat Shed/
Feed Storage

Mini-Farm

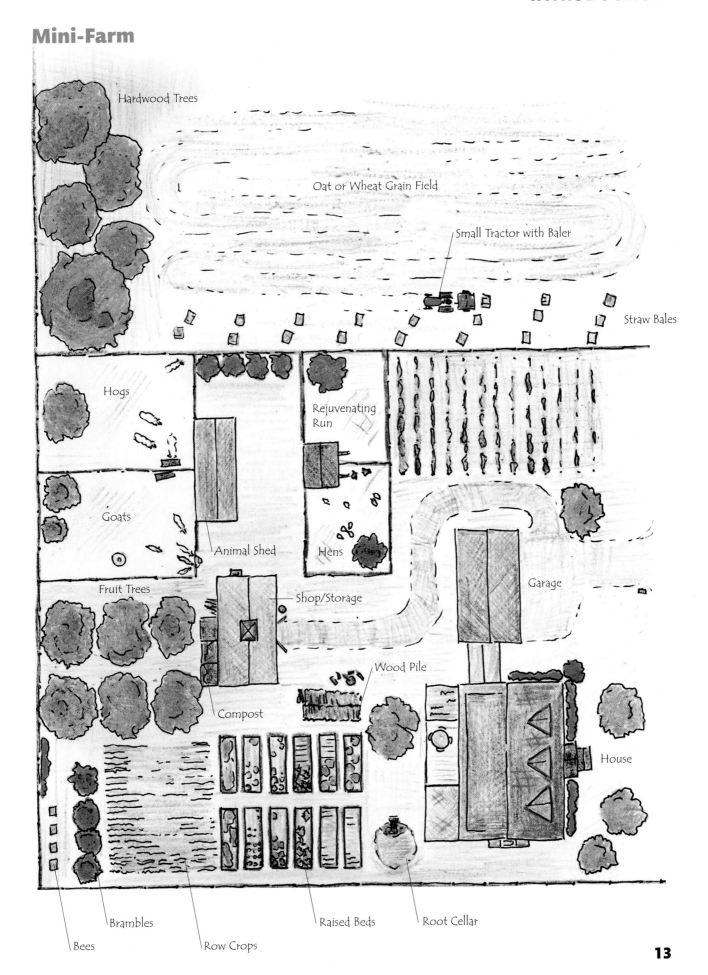

Hardwood Trees

Oat or Wheat Grain Field

Small Tractor with Baler

Straw Bales

Hogs

Rejuvenating Run

Goats

Animal Shed

Hens

Garage

Fruit Trees

Shop/Storage

Wood Pile

Compost

House

Brambles

Raised Beds

Root Cellar

Bees

Row Crops

1

Getting Started

Why Backyard Homesteading?

WHETHER YOU BEGIN with just a few tomato plants or plunge into a full-blown minifarm complete with chickens, bees, goats, and fruit trees, you'll find backyard homesteading is a life-changing experience. When you raise your own food, you know exactly what goes into it. You'll be assured that the produce you set on your table is free of herbicides, pesticides, and other questionable additives. You can enjoy vegetable varieties chosen for their great flavor, not their ability to withstand days in a semitrailer. You'll also have the opportunity to grow otherwise unobtainable heirloom vegetables, preserving valuable genetic stock while enjoying rare tastes and textures. And with better food readily available, you'll inevitably enjoy a healthier diet.

Besides having fresh food, you'll be stepping out of a supply chain that has been successful at providing food at low cost to the consumer but at a high cost to our soil, resources, and the animals that feed us. Instead of depleting soil, you'll actually be improving it. Instead of treating animals like units of production, you can raise them humanely. And by raising food at your doorstep, you won't require the fuel-intensive shipment of fresh vegetables and other products from commercial farms thousands of miles away.

Food Security

You'll also be able to provide for yourself and your loved ones, no matter what the outside world throws at you. If you have seen the aisles of empty supermarket shelves after even a couple days of winter blizzards, you know how tenuous our food chain is. Wise is the homeowner who has a bounty of preserved summer vegetables on the pantry shelves.

The ups and downs of personal finances are another good reason for starting a backyard homestead. If you've ever had to do some serious belt tightening because of a job loss, you know that the more you can do yourself, the less vulnerable you'll be. It's great to know you have a "Plan B," should you have to get by on a lot less income. Backyard farmers take delight in tallying the produce they harvest and then noting what they would have paid at the grocery or farmers' market. Even a couple of garden beds can bring in hundreds of dollars of produce each season.

Fresh vegetables from your own land, grown purely, are just one of the benefits of a backyard homestead. By raising them yourself, you know exactly what went into them.

A Fine Tradition

The lure of self-sufficiency has a long tradition. Nineteenth-century homesteaders were guided by how-to publications like *The Cultivator* and *The Prairie Farmer*. The *"Have-More" Plan*, published in the 1940s, led a generation of city dwellers back to a life of landed independence. In the 1960s, the *Whole Earth Catalog* offered access to tools and ideas about creative self-sufficiency.

Healthy Lifestyle

Many of us spend too much time at a computer all day, leading an indoor, sedentary life. A backyard homestead engages you in productive, healthy work devoid of the artificiality of driving to the gym and mounting a machine for a workout. Some weeding, cultivating, or pruning is often just what the doctor ordered for getting outdoors and breaking a sweat. And what a pleasure to return home from work and check the garden! You'll find it, too, has been busy and has new treasure to show off at day's end.

Kids also benefit from being involved in producing food for the family. It's too much to expect that they will love every chore, but honest work and knowledge of where food comes from are definite benefits.

Kids' enthusiasm for chores will ebb and flow, but by participating in your farming enterprise your children will be equipped with skills to someday produce their own food. Most importantly, they'll gain an indelible insight into where food comes from.

Is Backyard Homesteading Right for You?

It's easy to cite the benefits of becoming more self-sufficient in your food production, but a few cautionary words are worth considering. The first involves money. While a garden will save you money, other items will be more costly than if purchased. For example, the eggs you get from a small flock of laying hens will inevitably cost more than the factory-farmed variety from the grocery store. (Your hens will be much happier, however.) And a couple of gallons of milk from your goats simply can't compete with the cost of cow's milk from the mega market. The benefits are not in cost savings, but in freshness, purity, and your food independence.

You may also find yourself tied to the land in ways you didn't expect. The traditional two-week vacation takes a lot of arranging if you have livestock or a garden needing weeding and cultivating. With a little prep, chickens can get by on their own for a couple of days, but beyond that you'll need help while you are away. Dairy goats must be milked daily and are wily enough to require someone who knows goats to care for them. Many backyard farmers find compatriots with a similar enterprise with whom they can exchange caretaking.

Successful backyard farming takes experience. Be prepared for some failures. For example, planting times vary by region and by plant variety. Chance it too early, and the frost will nip your seedling. Sow too late, and the plant may not reach fruition before the end of the season. The quantity of vegetables you'll want also takes some experience. Everyone loves tomatoes, but too many plants will leave you buried. Animals are a study in themselves. Goats, for example, are frightfully intelligent and quite capable of breaking into the feed store and eating themselves sick. (See page 160.) Laying hens can go broody and refuse to budge from their eggs. You'll suffer a painful peck or two before you learn how to extricate the egg.

Learning the Ropes

Almost anyone can learn the skills necessary for backyard homesteading, but they'll come more quickly with the help of a mentor or two, especially where animals are involved. For example, beekeeping is not difficult but has subtleties that only an experienced hand can communicate. Local groups such as a bee club are a helpful source of knowledge, especially regarding local practices and conditions. Or if you are interested in keeping a couple of goats, observing milking or hoof trimming firsthand will jump-start your skills.

The Right Stuff

Getting in touch with the earth involves some earthy activities. Animals inevitably mean manure, though chicken and goat manure is benign stuff. (Hog manure is not.) And although small-scale farming is far less prone to such problems than are intensive farm operations, your animals may fall prey to pests and diseases. You will deal with the eventual death of an animal that, if not a pet, is at very least a familiar co-worker. One backyard homesteader opted to stop naming his chickens and give them numbers instead—a way of lessening the emotional tug.

And of course, if you raise animals for meat, you'll have to learn killing and butchering techniques—a job not for the fainthearted. Both are skills best learned under the guidance of someone experienced to avoid messy and potentially traumatizing mistakes.

Launching a garden in the spring involves intensive labor, especially as you rush to break ground and plant your first season. Rest assured it gets easier. An established garden requires less work as time goes on. But it does take maintenance: if a garden is going to be productive, don't expect to just throw the seeds in the ground and come back to pick the tomatoes. Weeding, mulching, pruning, watering, feeding, and staking are just a few of the essential chores. That's why starting small and growing incrementally is a good idea. Too ambitious a farmstead could bury you before you've learned the tricks of the trade and can work efficiently.

Trimming hooves is one of the many rough-and-ready chores you'll have to handle if you raise goats. (See page 167.) With any livestock you'll learn to deal with diseases, parasitic pests, and the inevitable reality of manure.

It's not surprising that gardening involves some earthy chores. Prepping compost for the coming season is necessary to ensure the productivity and health of your soil.

Municipal Regulations

Two types of municipal regulations relate to backyard farming. **Zoning** regulations concern changes that may infringe on a neighbor or could affect neighborhood property values. If you plan on having livestock, check local regulations on keeping animals, including requirements for the nature and location of their housing. (See pages 121–122 for specifics.)

If you plan on building structures, you'll need to determine local **setback** requirements. A setback is the distance from the lot line to a permanent structure such as a garage, home addition, porch, or deck. A typical setback is 12 feet from the lot line. It often does not pertain to temporary structures such as hutches, coops, and garden sheds—anything that could be dragged to different location. However, a permanent minibarn or shed will likely have to be positioned to allow for setbacks.

The curbside strip along the street is productive ground that often goes unused. Your municipality has the right to excavate there. In the meantime, it's a great place for raised beds. Keeping the beds in good shape and well tended is a neighborly courtesy.

Easements defend access to utilities and might include a neighbor's right to share your driveway, curbside strips (the area between the street curbing and sidewalks), or a public-access pathway running along your property. Choosing to plant crops in an easement area is perfectly reasonable if it is unlikely to be used soon. For example, a municipality has the right to dig up a curbside strip to reach water and sewage pipes, but routine system upgrades are planned long in advance. A visit to city hall will tell you whether any are planned for your parkway soon.

You may also have to abide by **Lot Coverage Ratio (LCR),** the proportion of your lot occupied by buildings and paving. And you'll have to follow the **building codes** that set local building standards. (See box below.)

Why Building Codes?

Building codes exist for your health, safety, and welfare. If you plan a permanent structure, pay attention to code requirements. Most are available from your building department via the Internet. Codes are typically based on sound engineering practices, setting requirements for things you otherwise might not think of—the effects of snow load, the need for footings that extend beneath the frost line, earthquake bracing, and the safest way to bring electricity to your structure. Permit costs vary, but often begin with a fee of $50 or so for the first $1,000 of the value of the shed or barn, plus about $10 per $1,000 of additional valuation.

In addition to zoning and building codes, there are intangibles to consider. Almost everyone is concerned about property values these days. The farming venture that may be the apple of your eye might be an eyesore for your neighbor. Be a good neighbor by planning your farmstead well, designing structures so that they are good looking as well as functional, and keeping things neat and in good repair.

Making a Plan

A master plan is the best way to make efficient use of your property. Even though it is unlikely you'll get everything up and running in your first year, a plan can help you avoid labor-intensive and costly mistakes. For example, it's no fun to build a neat series of raised beds only to later discover that you need to trench water and electricity lines through them. And if you locate the chicken coop away from intense sun, you won't later have heat-stressed and unproductive layers. Here are some planning methods for thinking through the makeup of your backyard operation.

Draw a Site Plan

Start with a site plan, a simple drawing that includes all lot lines, buildings, walkways, and fences on your lot. A plat survey is a handy beginning point—you'll find it attached to your mortgage documents. It will have many of the measurements you need, though you'll have to add measurements of your own and take into consideration any

improvements added after the plat survey was completed. Also sketch in predominant trees and bushes, and any paved or boggy areas. Ideally, complete your site drawing in the autumn. That way, you'll have the winter to mull over the possibilities before breaking ground in the spring.

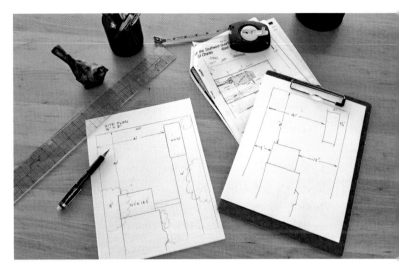

A site plan with rough dimensions for your lot is the starting point for a homestead plan.

Assessing Sunlight

Fruit and vegetables will be key components of your farmstead; both require sunlight. Before positioning anything else in your plan, make sure your vegetable beds, berry bushes, and productive trees are situated where they'll get adequate sunlight per day. You'll also want to be sure hutches and coops have adequate shade.

"Full-sun" plants—heavy producers like tomatoes, squash, peppers, and peas—need a minimum of six hours of sunlight. Any less, and they won't grow to full size. Such plants will do well with much more light than six hours, be it full sunlight or dappled shade. Other plants, like lettuce, some herbs, and cooking greens, do well in "partial shade"—about four hours of sun.

If you have no nearby trees or buildings that might shade the garden area, you're on safe ground. If your plots get varied amounts of sunlight you'll want to posi-

tion growing beds to take best advantage of available light. Make several photocopies of the table on page 20, and record sun patterns in various areas of your yard.

If you don't have all day to do the checks, buy a reusable sunlight meter. Some types stick on the ground, producing a reading after 12 hours. They'll assess whether the area is full sun, partial sun, partial shade, or full shade. You may need to increase sunlight in an area by removing trees or bushes. To preserve shading trees, try some creative pruning. "Limbing up" a tree means removing the lower branches to reduce the lower leaf canopy and allowing more light in.

Once you've determined sunlight patterns, sketch in any areas of questionable sunlight intensity for future reference. Also sketch in tree locations and a rough rendering of the shade they produce.

Sun Log

Photocopy this table and use it to assess potential crop areas. For a total survey, use it once a month throughout the growing season.

AREA

Hour	Full Sun	Partial Sun	Full Shade
6:00 a.m.			
6:30 a.m.			
7:00 a.m.			
7:30 a.m.			
8:00 a.m.			
8:30 a.m.			
9:00 a.m.			
9:30 a.m.			
10:00 a.m.			
10:30 a.m.			
11:00 a.m.			
11:30 a.m.			
12:00 noon			
1:00 p.m.			
1:30 p.m.			
2:00 p.m.			
2:30 p.m.			
3:00 p.m.			
3:30 p.m.			
4:00 p.m.			
4:30 p.m.			
5:00 p.m.			
5:30 p.m.			
6:00 p.m.			
6:30 p.m.			
7:00 p.m.			
7:30 p.m.			
8:00 p.m.			

Brainstorming

With a completed site plan in hand, you can begin the exciting business of deciding what your farmstead might include and where the elements might be located. One of the best brainstorming tools is a scale map of your yard on which you can place cutouts representing planting beds, sheds, coops, fruit trees—all the ingredients in your farmstead.

Quarter-inch graph paper is handiest for doing this. Large-format graph paper is available, but you can tape together enough 8½ x 11-inch sheets to represent the area you are planning. Transfer your site plan to the graph paper, including outlines of existing buildings, drives, and walkways. Then sketch in trees and areas of shade and full or partial sun.

Now the fun begins. Make scale cutouts of everything you'd like to see on your farmstead. If you plan to have a chicken coop or goat shed, work out in advance the general dimensions by consulting the relevant chapters in this book. Include dream stuff like a new pergola over the deck for grape vines, a jungle gym for the kids—anything you'd like to add. Get the whole household involved.

Take a digital photo of the each layout. (Lean over the center of the plan and shoot straight down.) Print out the options for comparison later.

Here are some things to consider as you plan:

- Well-cared-for livestock create little odor or noise, but to avoid potential problems, locate animal pens away from the lot line.
- You may be hauling in mulch, topsoil, lumber, and fencing, so include access pathways.
- Look for handy proximities. A series of compost bins between the chicken coop and goat shed could save a lot of hauling. An herb garden near the kitchen will get used more often than one at the back of the garden.
- Plan raised beds so that there is enough space to comfortably wheel your garden cart or wheelbarrow between them—about 2 feet. A raised bed 4 feet wide suits most people. Because the advantage of a raised bed is being able to weed and cultivate without stepping on the soil and thus compacting it, you'll want to be able to easily reach the center of the bed from the path. Length is limited only by available lumber.
- Fences should be set back from the property by at least 1 foot—sometimes more. Your municipality's Web site should have local specs.
- If you plan to keep bees, site the hives so that the flyway to honey and pollen gathering areas doesn't cross a neighbor's patio or other area where there may be a lot of people. A high fence lifts the flight path over people's heads—another way to sidestep a nuisance.
- Plan for water supply lines to service garden beds and livestock housing.
- Your chicken coop, garden shed, or shop area may need power. Code will dictate what type of cable is required and how deeply it will need to be trenched. (See page 28.)

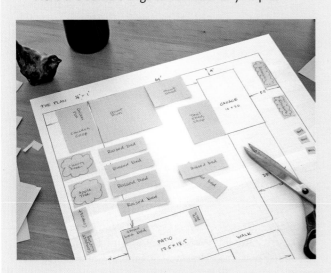

Scale cutouts of the major features of your homestead let you swap things around until you get an arrangement that takes full advantage of sunlight, has access paths, and locates things in convenient proximity.

Make a Final Drawing

By tracing around your cutouts, you can make a scale plan on a fresh sheet of gridded paper. You'll likely work out detailed garden-planting plans later, but now's the time to get specific about the location of fruit trees and brambles. For quick reference when working outdoors, write in as many dimensions as will be helpful once you start installing things—it's no fun counting ¼-inch squares every time you have to lay out a planting bed. Use this plan to work out underground water and power lines. Once you've decided where they'll run, add them to the plan.

Make a Model

Foam-core board is an ideal material for making a ½-inch-to-the-foot scale model of your future farmstead. Some boards come with a preprinted ½-inch grid. While modeling may seem like overkill, many people find three dimensions much easier to visualize than flat drawings. In a couple of evenings you can make nifty miniature sheds, raised beds, beehives—all the components. A model is also a great way to get kids involved. Moving the pieces around will capture their imagination and get them looking forward to the project.

Downloading and printing a ½-inch grid and taping sheets together will give you scaled base for your model. You can then slide things around to see what arrangement works best. This method can also help you decide on fence heights, especially if you want to screen a view or buffer prevailing wind.

Make a working drawing of your final plan, including dimensions for key items. Load it up with any notes that may prove handy while you are working outdoors.

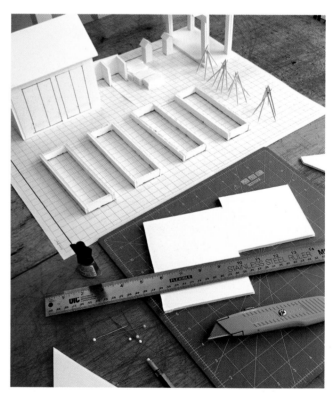

In an afternoon, you can make foam-core components of your farmstead for a 3-D view of how things might look. Modeling often captures the imagination of kids—a great way to get them involved in the process.

Reality Check

As a final check on your plans, go outside and mark the location of major items on your plan. Stretch lines, and use landscaping spray paint or baking flour to mark the footprints of sheds, fences, and beds. You'll get a real sense of how things will go together. Very likely you'll spot some glitches you hadn't anticipated.

Planning for Plants

The most productive part of your farmstead will inevitably be the things you plant. If your space is limited, consider the potential usefulness of every inch of your property. Here are some possibilities:

■ **Plant the Curbside Strip.** This otherwise wasted space can be quickly transformed into attractive, productive beds. (See page 18.)

■ **Plant your Front Yard with Attractive and Productive Fruit Trees.** In many cases, these plantings are just as good looking as their purely ornamental cousins.

■ **Add Raised Beds.** They're easy to work, productive, and space saving because you can position plants closer together.

■ **Use Containers.** Set them on stair steps, decks, porches, and patios; they put odd corners of your lot to work.

■ **Double-Dig Your Plots.** (See page 39.) Your plants will weather drought better, and you'll squeeze more produce out of limited space.

■ **Combine Climbing Crops with Spreading Crops.** For example, spreading squash and pumpkins can go between climbing pole beans.

■ **Train Spreading Crops to Go Vertical.** Cucumbers and squash can grow up trellises, saving space and preserving the harvest from insect damage.

Corn reaches its full height late in the season and can otherwise be a space waster. Planting squash between the corn rows keeps down weeds and frees up ground for other uses.

Scale

Small gardens tend to be neat gardens. That's because it is easy to keep on top of a small garden: weeds get pulled when they are tiny; and crops are harvested when they are ripe. If a disease or insect threatens a plant, it's easy to notice it before it gets out of hand. If you build the soil with compost, mulch to keep down the weeds, and plant intensively, small vegetable gardens offer surprising yields.

If you are a beginning vegetable gardener, start with a garden that ranges between 15 x 20 feet and 20 x 30 feet—anywhere from 6 to 12 raised beds 4 feet x 12 feet. A 15- x 20-foot garden is large enough to grow fresh vegetables to feed a family of four as well as a few extras for preserving or storing for winter use. In a 20- x 30-foot garden, you can grow some of the space hogs such as sweet corn and pumpkins, plus summer vegetables and a few preserving crops.

If you are new to gardening, start small and expand as you learn the ropes. It pays to learn what size plot you can handle; start too large, and you might get overwhelmed and discouraged.

As you gain familiarity with each crop's needs, build your soil, and learn how to deal with pest and disease problems, it will be easy to increase the size of the vegetable garden.

Planning for Fruit Trees and Brambles

Choosing where your fruit trees are to be situated is as important as deciding what species and cultivars to grow. If possible, take a year to make this decision. During that time, you'll be able to observe various locations closely. You will know where frost comes first and last, where puddles form in heavy rainstorms, and where winds are strong, moderated, or stagnant. In cases where you simply do not have time to learn about the site, try to take the following characteristics into consideration.

Fruiting shrubs are some of the most beautiful and useful plantings in your yard. When you plan your farmstead, consider whether a fruiting shrub can supply the necessary form or color required for the design. For example, if you need a high hedge to buffer highway noises, try growing high-bush blueberries or elderberries. Prickly shrubs such as gooseberries can make a hedge that keeps deer and dogs from entering the yard, and half-high blueberries are wonderful foundation plantings in combination with some evergreen plants.

No matter where you put your brambles, they will grow large enough to become a visual focal point. The only way to make them unobtrusive is to hide them behind a building or hedge. But inasmuch as this isn't always pos-

sible, try to find a place where you can use their growth habit to your advantage. They can define an area of the yard or separate one spot from another. A small planting can even grow between seedlings of standard-size trees.

Fruit trees and berry bushes take years to come into full production, so carefully assess their location to avoid cold spots, winds, and dampness. Consider the front yard, which is an ideal place to combine beauty with productivity.

Create an Edible Landscape

We tend to think plants are either decorative or edible, but in fact many are both. Nasturtiums have lovely, edible flowers. Kale is a tasty table vegetable whose bold purples are an asset to any border. Beds of strawberries are beautiful. Very likely most of the plantings in your unimproved yard are solely decorative. Establishing a productive yard doesn't call for pulling out anything that doesn't produce food. Instead, slowly add edible plants and see how they fare. Aim toward a blend that looks good and earns its keep.

Even a highly productive garden can look beautiful. Adding an arbor with climbing plants and including attractive fencing makes a garden a feast for the eye as well as for the table.

Rooftop Gardening

Finding enough sunlight and space are your two major challenges if you are growing crops in the city. With access to a rooftop, those challenges pretty well evaporate as long as you can deal with one fundamental difficulty, a garden's intrinsic weight.

Dirt is Heavy. Wet dirt is even heavier. To simply build a bunch of raised beds and load them up with soil could bring the roof down with the first heavy rain. The surest way to determine what a roof can bear is to hire a structural engineer to assess it. In any case, you'll want to lighten the soil as much as possible, mixing in materials like pumice, coconut fiber, peat moss, perlite, expanded slate (such as Stalite PermaTill, which is one commercial product designed for gardens), or vermiculite. It's possible to blend these materials with as little as 25 percent compost to come up with suitable rooftop soil.

Containers should also be as light as possible. Some rooftop gardeners get creative with old dresser drawers, fruit crates, and even broken-out guitars. Wood raised beds work as well, though it's best to use 1-by instead

Urban Backyard Crops

The bane of urban gardening is shade. There are just too many buildings around to let in the sunlight a garden needs. The solution is to be very selective about where you locate plants. Sun-hungry plants such as tomatoes need to go where they'll get six or more hours of sunlight. Root vegetables and greens can go in partial shade. Containers on landings, stairs, and balconies help capture sun, as do any vertical or overhead arrangements such as trellises, arbors, and stacked planters. Window greenhouses are a great way to save outdoor space for other produce and keep herbs and lettuces close to the kitchen. Community gardens on vacant lots or city-owned property are often the best way to supplement backyard gardens.

of 2-by material to conserve weight. Large plastic pans manufactured especially for rooftop gardens (GreenGrid is one brand), make a neat installation. Plastic kiddie pools are a colorful option and work well because they add almost no weight, are water tight, and hold enough soil that they won't blow away in heavy winds.

Wind is an important consideration. Rooftops get plenty. Secure small containers, and always have some means for keeping garden tools from blowing off the roof. Also be sure to secure any plants that climb trellises or stakes. The other effect of wind is dehydration. Coupled with intense sunlight, it can quickly dry out moisture in the soil. Water frequently or, better yet, install a microirrigation system. (See page 27.) Your home center has inexpensive systems with battery-operated timers that will keep your beds well irrigated.

Rooftops offer productive space for raised-bed gardens. Mixing dry matter like pumice, coconut fiber, peat moss, and expanded slate keeps soil light enough so that it doesn't overburden the roof. This garden atop the Burnside Rocket building in Portland uses kiddie pools and supplies produce for a restaurant called Noble Rot.

Water

If you have a few chickens, you can get by with a garden hose to replenish their water, but if you have goats as well, you'll need a handy water source. In addition, if you plan on a lot of crops, you'll need a water supply for micro- or macro-irrigation. Electricity might also seem like a luxury, but you'll need it for outdoor lighting, power for heaters to keep stock waterers from freezing, and for extending daylight hours to keep laying hens productive. Make your plans for water and electricity supply lines early on. Both require trenching, something you'll want to finish before installing compost bins, building raised beds, or putting up any coops or sheds.

Water Supply

A water line can be run from the main supply in your house as long as it has two important features. First, it must have an antisiphon valve to keep contaminated water from being sucked into your home's water system. Second, it must be drainable to keep water from freezing in the pipes during the winter and causing damage, be set up so it can be blown out with compressed air each fall, or be completely winterized to eliminate freezing.

It's always best to bury water lines so that they don't get damaged when digging, but if you need water through the winter, the lines must be buried at least 6 inches below the frost line to keep from freezing. They'll also need heat tape where tubing is exposed aboveground. Water lines can be flexible hose (though some critters may chew through this), PVC, or steel. Before digging, contact your local utility or one of the utility-locator services, such as "Dig Safe," "One Call," or "Julie," to be sure you won't be trenching into existing utility lines. Map out your underground pipe once it is installed so that you won't dig into it later.

An antisiphon valve is essential equipment for keeping potentially tainted water from flowing back from the garden or livestock area into the household system.

Gathering Rainwater

Collecting water from rooftops and holding it in a barrel seems like a great idea, but there are some limitations. Foremost is quantity. The biggest barrels hold just over 100 gallons, enough to get you by for a few days in drought conditions. In addition, some people are concerned the heavy metals shed by composite roofing may contaminate their vegetables. Bottom line: a rain barrel is good idea for supplemental water that would otherwise be wasted, but a large garden can't depend on it for an extended dry spell. If you use one, be sure it is covered to keep it from becoming a mosquito nursery and has some sort of overflow setup.

Microirrigation Systems

Getting the right amount of water exactly where it is needed saves water, conserves energy, and helps your garden flourish. Hand watering with a hose keeps you in touch with your plants, but in the long run, it's laborious and wasteful. Simple microirrigation systems run off a valve timer attached to a hose spigot. From that you can devise a system that, with a bit of tinkering, will put just the right amount of water at the base of every plant in your garden. Microirrigation is also ideal for containers that are otherwise prone to dry out quickly. Even hanging containers are easy to connect to the system.

You'll like working with your microirrigation system. The tiny spray heads, some of which rotate, and emitters are darn cute—almost like playing with building blocks. The best bet is to buy a starter kit with all the necessary components, and add from there. Essentially, a microirrigation system comprises ½-inch feed hose, to which ¼-inch tubing is attached with punch-in connectors. ("Goof plugs" seal mistakes or altered configurations.) The tubing runs to the sprayers or emitters,

some of which are placed on plastic stakes called risers where an elevated spray is needed. Systems typically run early in the morning and late in the evening to reduce evaporation. You'll get a kick out of hearing the battery-operated valve click open and watching the sparkle of water throughout your crops.

A weep line works well for crops like lettuce that are densely planted. By directing water toward the base of the plant, there is no overspray that can cause damaging burns in full sunlight.

Microirrigation Components

Microirrigation gets the right amount of water exactly when and where the plant needs it. Waste and evaporation is minimal. Assembly is almost fun.

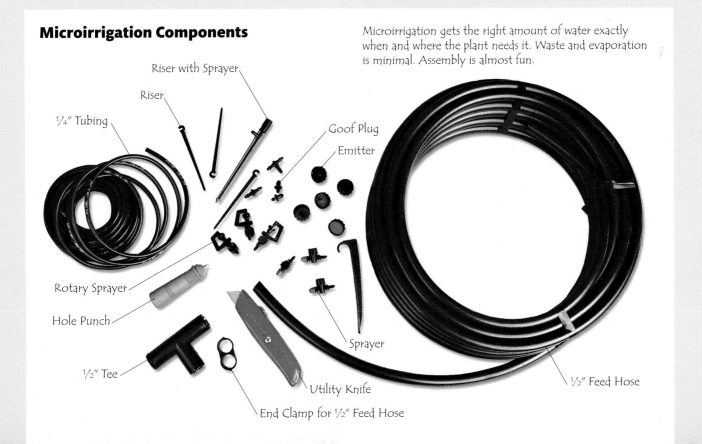

¼″ Tubing

Riser

Riser with Sprayer

Goof Plug

Emitter

Rotary Sprayer

Hole Punch

½″ Tee

Utility Knife

Sprayer

End Clamp for ½″ Feed Hose

½″ Feed Hose

Electricity

An electrical supply line can be run off your main breaker panel by having an electrician install a new circuit—usually a simple procedure. Any outdoor receptacle should be protected by a ground-fault circuit interrupter (GFCI). Or you can install a GFCI-breaker to protect the entire circuit. Unless you live in an unincorporated area, you are required to bury the electrical line. Burying the line is a good idea under any circumstances to avoid power outage due to a downed line. Use only underground feed (UF) cable. To protect the line, pull it through conduit. (Avoid putting power and water lines in the same trench; use side-by-side trenches instead.) Save money by digging the trench yourself, but be aware that depth is measured from the top of the conduit, not the bottom of the trench. You'll need an inspection before burying the conduit. If you have outbuildings, you might want to install a three-way switch in the house so power to them can be controlled remotely—it'll save you from having to run outside to turn off lights left on by mistake.

Lighting

Lighting may seem extraneous, but stuff happens—especially when you keep livestock and have a nighttime emergency with which to deal. Even if you only have vegetables, you'll very likely have a late evening when you have to cover tender spring plants threatened by frost. Lighting is much easier to install early in the game when you can trench where you please. Here are some situations that may require lighting.

- **A nighttime check of the henhouse or goat shed is always a good idea, especially in cold weather when you'll want to be sure water is available. You'll want the fixture on a timer if you plan to extend daylight to keep the eggs coming.**
- **A motion light on a shed or overlooking the garden often proves handy. Solar powered LEDs save you the trouble of an electrical hookup.**
- **If you dedicate part of a shed to a shop area, you'll want plenty of lighting. If it is unheated, you won't be able to use standard fluorescents. Instead, install cold cathode compact fluorescent lightbulbs that will work in temperatures down to -10° F.**

Use only underground feed (UF) cable if you bury your electrical line. Most codes require it to be 6 inches below the frost line so that the line won't be damaged by frost heave. If your area permits a shallow trench, conduit is a prudent safeguard against accidentally damaging the cable while digging.

Underground conduit is not always required but is a good safeguard from unintentional damage. Wire lubricant eases pulling UF cable through the conduit. Always check local codes for outdoor wiring requirements.

A GFCI receptacle protects you from shocks—a necessary feature for outdoor use. One GFCI receptacle can protect a series of standard receptacles downstream from it. Another option is installing a GFCI breaker in your main panel to protect the entire outdoor circuit.

Permission to Buy Tools

Here are the essential tools you'll need for general gardening and livestock chores. If you are starting on a tight budget, search garage sales and flea markets. You'll find perfectly good tools that will give you years of service.

Shovel Definitely one of the essentials for digging, scooping, and filling

Hoe Essential for clod busting, furrow making, and general cultivation

Garden Fork Necessary for double digging and turning compost; can substitute for a hay fork in a pinch

Metal Rake Intended for dealing with soil, not leaves; pulls vegetation out of freshly turned soil, breaks up small clods, and smoothes dirt

Scoop Handy for dealing with manure, gravel, and sand

Cultivator Perfect for aerating the soil and teasing out weeds

Hay Fork Ideal for laying down straw bedding and mulch

Lawn Rake Great for gathering leaves and dead vegetation

Kids' Tools Recommended: functional miniatures of the real thing; not recommended: plastic tools that don't really work

Baby Sledge Very helpful for pounding in stakes and poles

Bypass Pruner If you have fruit trees, great for pruning branches up to ¾ in. in diameter

Trowel Essential for planting sets

Hand Cultivator Great for breaking up the soil and removing weeds that grow close to plants

Bow Saw Indispensable for pruning anything the bypass pruner can't handle and dealing with fallen limbs and other rough cutting chores

Tool Maintenance

Sharp, clean tools make the job easier. Before each gardening season and after heavy use, use a bench grinder to give tools a new edge. Keep your garden tool clean and rust-free by plunging it into a bucket of coarse sand after each use.

Small Shed

This handsome 8 x 10-foot structure makes a roomy garden shed, small shop, or a first-class chicken coop. With a ridgeline that rises 14 feet above the ground, you can add a roomy loft for storage. The 80 square feet of space below leaves ample room for garden tools, a bench, and feed storage. Its modest size and robust floor framing make it an ideal shed for a block foundation. Lay down landscaping fabric and gravel first. (See page 33 for more foundation options.)

Because two sides of the shed are 8 feet long and framed with studs that are 16 inches on center, you can use full sheets of plywood for sheathing. For the two 10-foot walls, 8-foot lengths of plywood will cover all but 2 feet of the wall; cutting a single sheet of plywood into four pieces makes up the difference. The gable roof rises 4 feet above the wall's horizontal top plates, so a single sheet of plywood, notched for the exposed ridge beam, covers each gable end.

The cedar shingles shown are available in individual shingles or, at a greater cost, prefabbed as single-course panels. More affordable is exterior-grade plywood with 1 x 2 battens every foot to cover seams and achieve a barnlike look.

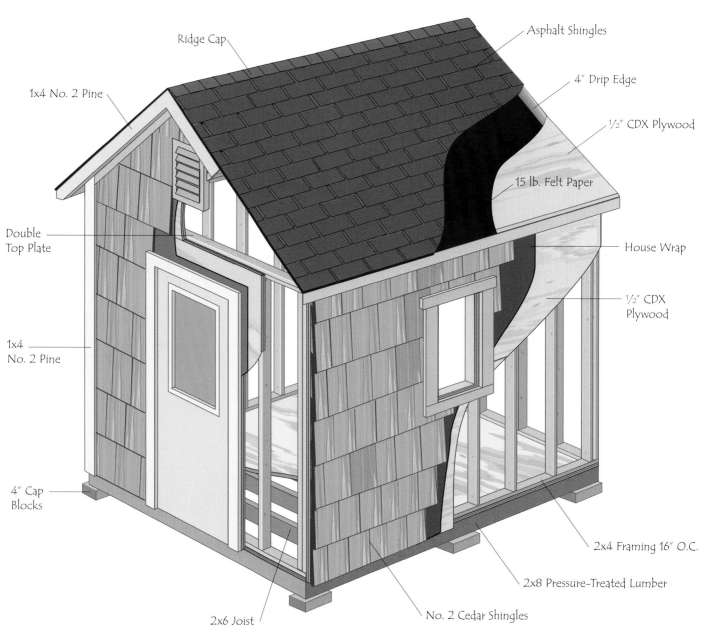

Ridge Cap

Asphalt Shingles

1x4 No. 2 Pine

4" Drip Edge

½" CDX Plywood

15 lb. Felt Paper

Double Top Plate

House Wrap

½" CDX Plywood

1x4 No. 2 Pine

4" Cap Blocks

2x4 Framing 16" O.C.

2x8 Pressure-Treated Lumber

2x6 Joist

No. 2 Cedar Shingles

Shed Construction Tips

- Sheds offer an ideal opportunity for using salvaged and recycled lumber.
- Use pressure-treated lumber for any part of a structure that comes in contact with soil or within one foot of grade.
- Prefab as much as possible. Frame and sheath walls before raising them. If you build a door, attach it to its jamb (the frame surrounding it) before mounting it in the rough opening of a wall.
- If are new to building, consider using deck screws for the framing. Not only do they hold better than nails, but if you make a mistake you can easily back them out. If you use two cordless drills, one for pilot holes and another for fastening, you'll approach the speed of nailing.
- Use only galvanized or coated fasteners on the outside of the shed.
- If the shed will be used as a shop, position windows low enough to run long boards through them when using a radial arm saw.
- Plan water supply lines before building. Even if you don't need water in the structure, it may prove best to attach a hose spigot to the outside of the shed.

Materials List

Foundation

3	50-lb. bags of ¾" gravel
18	4" cap blocks
2 pcs.	2x8 10' pressure-treated side rim joists
2 pcs.	2x4 10' pressure treated
10 pcs.	2x6 8' pressure treated
12 tubes	Construction adhesive

Framing and Sheathing

15 sheets	½" CDX plywood
42 pcs.	2x4 8' SPF
16 pcs.	2x6 8' SPF
1 pc.	2x8 12' SPF
2 louver vents	16" x 20"
1 roll	house wrap
10d common nails	
8d common nails	
2½" coated screws	

Roofing

8 pcs.	4" drip edge
1	100' roll #15 roofing felt
6 bundles	Packs of 3-tab shingles
⅞" roofing nails	

Trim and Siding

17 pcs.	1x4 8' No. 2 pine
10 bundles	No. 2 cedar shingles
2½" galvanized finishing nails	
1 gal.	Cedar sealer
1 gal.	Primer for trim
1 gal.	Paint for trim

Doors and Windows

1 prehung door	
2 36" x 40" windows	
1 pack cedar shims	

Stairs

6 pcs.	4" cap blocks
2 pcs.	2x6 10' pressure treated
10 pcs.	2x8 8' pressure treated

Large Shed

At 12 x 16 feet, this shed is big enough for a shop area (a necessity veteran backyard farmers swear by), feed storage, and a full complement of gardening tools, with room left over for storing a tiller or small tractor. The windows let in plenty of light, and the double doors make it easy to move awkward items in and out. The classic Saltbox design suits most any architectural style, so it's bound to fit in with your home.

The walls are framed with 2x4s, with ½-inch plywood or OSB sheathing to make a solid shell. Siding can vary from the beveled siding shown to less-expensive T1-11 plywood siding or cement-board panels. The painted wood trim is ⅝ pine.

Framing the roof and eaves the biggest challenge. Work from rented or borrowed scaffolding to make the job easier. Roll roofing is the least expensive roofing option but will serve for only 10–15 years. Composite shingles are good for 20–25 years.

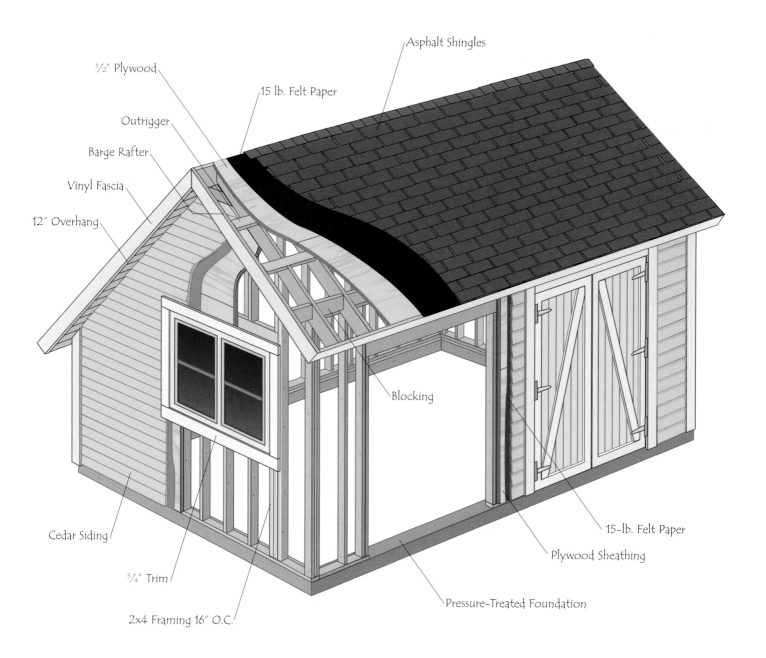

- ½" Plywood
- 15 lb. Felt Paper
- Asphalt Shingles
- Outrigger
- Barge Rafter
- Vinyl Fascia
- 12" Overhang
- Blocking
- Cedar Siding
- ⅝" Trim
- 2x4 Framing 16" O.C.
- Pressure-Treated Foundation
- Plywood Sheathing
- 15-lb. Felt Paper

Foundation and Floor

The simplest foundation for this shed is 6x6 pressure-treated timbers set on a gravel footing, with a floor of compacted screenings. When wet down, the screenings compact to form a smooth, hard surface. For a more finished floor, build a frame of 2x6s 16 inches on center (O.C.) supported by blocks every 4 feet.

Before building your shed, confirm that the site you've chosen is in compliance with local setback requirements. In the event you are close to compliance but not quite there, apply for a vari-ance. Submit it to the local zoning board or building department.

Other foundations: for small sheds, skids or stacked block are adequate, but for a shop you may want to start with a concrete slab. Even with skids or blocks, level the area in advance; lay down landscaping fabric; and cover it with about 6 inches of gravel for adequate drainage. Concrete piers are a step up from blocks, requiring only postholes dug to beneath the frost line. To raise the pier above grade level, add a fiber form tube before filling the hole with concrete.

Materials List

Foundation

8–14 pcs.	6x6 16'
8–14 pcs.	6x6 12'
84–150 spikes	10" galvanized
20 rebar spikes	24"
Gravel, screenings	

Framing and Sheathing

28 sheets	½" CDX plywood
2 pcs.	2x10 10'
19 pcs.	2x8 12'
19 pcs.	2x8 8'
3 pcs.	2x6 12'
7 pcs.	2x4 16'
5 pcs.	2x4 14'
12 pcs.	2x4 12'
28 pcs.	2x4 10'
11 pcs.	2x4 8'
4 pcs.	1x8 10' No. 2 pine
8d, 12d common nails	

Roofing

9 pcs.	Aluminum drip edge, 10'
3 rolls	15-lb. builder's felt
13 bundles	Shingles (3 bundles per square)
⅞" roofing nails	
⅝" staples	
2 tubes asphalt cement	

Windows

4 34" x 36" double-hung windows

Trim Boards

15 pcs.	⅝x6 12' No. 2 pine
6 pcs.	⅝x6 14' No. 2 pine
6 pcs.	1x6 12' No. 2 pine
6 pcs.	1x8 10' No. 2 pine
12d finishing nails	
3 tubes latex caulk	
1 gal. latex exterior primer	
1 gal. latex exterior paint	

Soffits and Fascia

13 pcs.	Vinyl soffit, 12' long
8 pcs.	Vinyl fascia, 12' long
8 pcs.	Vinyl J-channel
Aluminum trim nails	

Siding

1060 lin. ft.	½" x 8" western red cedar beveled siding
2" aluminum siding nails	
1¼" stainless-steel screws	
1 gal. latex semitransparent exterior stain	

Doors

4 sheets	¾" exterior birch plywood
13 pcs.	1×6 8' No. 2 pine
1 pc.	1×8 8' No. 2 pine
2 tubes construction adhesive	
6d finishing nails	
12 T-hinges	
48 screws 3" No. 12	
48 hex bolts w/washers, split washers, and nuts, 5/16" x 2"	
2 clasps and staples w/mounting screws	
2 sliding bolts w/mounting screws	

2

Raising Vegetables and Herbs

A VEGETABLE PATCH WILL BE THE CENTERPIECE of your farmstead, producing bushels of fresh, wholesome food with only a modest amount of effort on your part. By growing your own, you know exactly what went into your food, and you get to enjoy a degree of freshness and variety almost unattainable otherwise. You'll also find that you have saved a bundle compared with what you'd pay at the grocery store or farmers' market. Added to that, you'll find gardening is great exercise and an ideal antidote to the computer screen many of us are chained to in our working life.

Deciding What to Grow

Begin choosing what to grow by making a list of the vegetables that your family eats most frequently and the approximate number of heads or pounds that you use each week. The table "How Much Should You Plant?" (on page 36) lists approximate amounts of each vegetable that the average adult eats each year and the space required to produce that quantity, figured on both a row-foot and a bed-foot basis.

Next, order some seed catalogs. You can find all the information you need on-line. (See Resources, page 238 for seed suppliers.), However, you might enjoy engaging in the fine old tradition of perusing seed catalogs on cold winter nights. Make a point of including catalogs from seed suppliers located in your region. As you read, note disease resistance and seasonal or cultural recommendations. Try to steer clear of plant varieties that are said to "ripen uniformly." This means that the crop has been bred for the convenience of commercial growers; it may taste good, but the flavor is secondary to the harvest characteristics. Good seed catalogs guide you by letting you know what cultivars perform best in your area and in the different parts of the season. Make a list of your favorites, but hold off on ordering until you've made a garden plan. It's easy to get carried away.

Growing your own vegetables allows you to be nearly as choosy as you want to be. However, if you grow exactly the same thing in the same spot every year, certain weeds, pests, and diseases will build up. So if you want to grow nothing but Brussels sprouts (or sweet corn), you'll need to find four different places to grow your favorite crop so that you can "rotate," or change its location, each year for four years. If you grow a more varied assortment of crops, however, you can set up a rotation system that puts each crop in a different part of the garden over a four-year cycle.

Mapping the Garden

Map your vegetable garden so that you can place plants that are good companions adjacent to each other and keep poor companions far apart from each other. Consider successive plantings of each crop so that you'll have enough space in the right place when you need it. Make notes about planting depth and distances between plants to aid in planting. Save your maps from year to year so that you'll be able to rotate crops every year for at least four years.

How Much Should You Plant?

CROP	YIELD LBS. per row foot	LBS. NEEDED per adult	ROW FEET per adult	ROWS per 4'-wide bed	BED FEET (LENGTH) per adult
Asparagus	0.25–0.33	7	28–21	7	4–3
Beans, bush, snap	0.8	8	10	2	5
Beans, pole	1.5	10	6.67	2	3.34
Beet, greens	0.4	4	10	3	3.34
Beets	1	10	10	3	3.34
Broccoli	0.75	15	20	2	10
Brussels sprouts	0.6	5	8.34	2	4.17
Cabbage	1.5	12	8	2	4
Cantaloupe	1	10	10	1	10
Carrot	1	20	20	3	6.67
Cauliflower	0.9	10	11.11	2	5.56
Chard	0.75	6	8	4–5	2–1.6
Collard	0.75	6	8	3	2.67
Corn	0.96	30	31.25	2	15.63
Cucumber	1.2	10	8.34	1	8.34
Eggplant	0.75	10	13.34	2	6.67
Kale	0.75	5	6.67	2	3.34
Leek	1.5	6	4	3	1.34
Lettuce	0.5	30	60	3–7	20–8.6
Onion	1	20	20	3	6.67
Parsnip	0.75	10	13.34	3	4.45
Pea, English or shell	0.2	5	25	3–4	8.34–6.25
Pea, snow	2	6	3	3–4	1–0.75
Pepper	0.5	10	20	2	10
Potato	5	50	10	2	5
Rhubarb	0.8–1.2	8–12	6.67–15	1	6.67–15
Rutabaga	1.5	5	3.34	5–7	0.66–0.48
Salad greens, misc.	0.5	25	50	3–7	16.67–7.15
Spinach	0.75	8	10.67	3	3.56
Squash, summer	2	8	4	1	4
Squash, winter	2	20	10	1	10
Strawberry	1–3 (qts.)	30 (qts.)	30–10	2–3	15–3.34
Tomato	1.5	20	13.34	2	6.67
Turnip	1	8	8	5–7	1.6–1.15

Companion Planting

Plants exert strong influences over each other. They release chemical compounds from their roots and aboveground tissues that affect other plants as well as insects and microorganisms. Some of these effects are directly beneficial, as in the case of a substance exuded by the root that stimulates the growth or flavor of a nearby plant. Other companion planting effects are indirectly beneficial, as in the case of a flower that serves as a nectar source for an insect that preys on the pest insects of neighboring plants. Still other companion planting effects are negative, as in the case of a root exudant that inhibits the growth or flavor of another plant. The table "Companion Plant Influences," on page 38, lists the effects of combining vegetables with each other as well as some weeds, herbs, and flowering plants.

Succession Planting

Plan successive plantings right from the beginning. Short-season crops, such as lettuce, can follow other short-season crops. If the soil is very fertile, you can simply replant the area. But it's often best to apply at least ½ inch of compost, a light dusting of alfalfa or soybean meal, or a balanced organic bagged fertilizer for the second crop. Work the material into the top couple of inches of soil a week or so before direct-seeding or a few days before transplanting.

An alternative to composting is to grow a green manure (frost-tolerant annual grasses, grains, or legumes such as oats or fava beans, which can be planted in the early spring) until it is time to plant the crop. If you are planting later in the summer, use a tender crop such as buckwheat. Two weeks before planting the vegetables, till in the green manure.

Crop Rotation

Certain crops deplete or replenish particular nutrients in the soil, and some pests and diseases are drawn to soils where their favorite hosts have recently grown. By rotating your crops you can vary the soil nutrient requirements as well as minimize potential pests and diseases. It boils down to two essential principles. First, leave at least four years between planting members of the same plant family in the same spot. Second, follow heavy feeders, such as corn and squash, with light feeders like beets, beans, or peas. Specifically, here is a proven plant-by-plant rotation:

- Follow deep-rooted crops, such as broccoli, with shallow-rooted crops, such as onions.
- Potatoes follow sweet corn.
- Sweet corn follows the cabbage family.
- Cabbage family crops, undersown with legumes, follow peas.
- Peas follow tomatoes.
- Tomatoes, undersown with a nonhardy green oat manure, follow beans.
- Beans follow root crops.
- Root crops follow squash or potatoes.
- Squashes follow potatoes.

Timing

If you plant all of your lettuce in the first week of April, it will all mature at the same time. Rather than plant everything at once, experienced growers plant small amounts of each of the short-season crops all through the season. Most people take a few years to develop a "starting schedule" appropriate to their climate and chosen cultivars. The "Days to Maturity" listed in seed catalogs are fairly accu-rate for late-spring and early-summer plantings. However, temperature can greatly influence the speed with which vegetables mature.

With experience, you'll learn how long each crop will stand in the garden without losing quality. For example, the earliest spring lettuces and greens stay good for a week or two. But in midsummer, they become tough and bitter very quickly. Until you know your climate and cultivars, plan so that salad crops slightly overlap each other.

Companion Plant Influences

Some plants repel specific pests, and are good companions for plants that are attacked by them. "Trap" crops lure pests away from plants. Other plants produce a substance that chemically inhibits the growth of plants.

SPECIES	ATTRACTS	REPELS/INHIBITS	COMMON COMPANION CROPS
Allium (Onion, garlic, leek, scallion)		Aphids, peas	Roses, daffodils, tulips, aphid hosts in vegetable garden
Beans, bush		Onions	Carrots, cauliflower, beets, cucumbers, cabbage
Broccoli/cabbage		Tomato	Dill, celery, chamomile, sage, beets, onions, potatoes
Carrots		Dill	Onions, leeks, herbs, lettuce, peas
Catnip	Small beneficials	Catnip tea repels flea beetles	
Celery			Leeks, tomatoes, cauliflower, cabbage, bush beans
Chamomile	Small beneficials		Onions, cabbage family
Dill	Small beneficials	Carrot	Cabbage, beets, lettuce, onion
Eggplant	Potato bugs		Potatoes
Fennel		Many species	Plant by itself
Garlic		Aphids, Japanese beetles, mites, deer, rabbits	Roses and many vegetables
Hyssop	Bees	Cabbage moth, radish	Grapes, cabbage
Kohlrabi		Potatoes, beans	Onions, beets, cukes, cabbage family crops
Marigold		Nematodes, Mexican bean beetles	Tomatoes, beans, potatoes
Nasturtium	Aphids	Squash bugs	Trap crop with cabbage family
Nicotiana	Potato bugs		Potatoes
Orach		Potato	
Parsley	Small beneficials		Interplant with carrot, rose
Petunia		Mexican bean beetles	Beans
Potato		Repels Mexican bean beetles; inhibits tomato, squash, sunflower, raspberry	Beans, corn, cabbage, eggplant, horseradish, marigold
Rue		Japanese beetle	Inhibits many plants
Sage		Cabbage moth, carrot flies	Cabbage family, carrots
Sunflowers		Nitrogen-fixing bacteria, potatoes	Corn, cukes
Stinging Nettle	Small beneficials		Almost all plants benefit
Summer Savory		Mexican bean beetles	Beans
Thornapple		Japanese beetles	Near grapes, roses, pumpkin
Tomato		Asparagus beetle	Asparagus, gooseberries, roses

Preparing the Ground

New houses are often surrounded by bare ground. Sometimes the soil is so badly compacted that not much will grow in it. Older homes may have soil that has not been truly cultivated for years and is in need of rejuvenation. If the soil is overly compacted, try to get some air into it with a spading fork. After you've been over the whole area a couple of times with the fork, spread 1 to 2 inches of compost or fully aged manure over the whole planting area. Depending on the results of your soil test (page 42), adjust the pH and fertilize for nutrient deficiencies while you're

at it. Then sow a cover crop. If you plan to leave the cover crop in place for at least 18 months, include a nitrogen-contributing legume, such as alfalfa, in the mix.

Working with a Tiller

Tillers can make fast work of a heavy job, but don't till if the soil is wet. Not only is wet soil miserable to work, you risk creating clods that are impenetrable to roots as they grow. Evaluate your soil by grabbing a handful and squeezing. Consider it too wet to till if beads of moisture form on its surface or if it does not immediately break apart when you strike a handful with the edge of your other hand.

Mulching

Sheet mulching is one of the easiest ways to create a new garden area in a lawn, and it contributes enough nutrition to make up for all but the most serious soil deficiencies. Sheet mulching involves laying down a weed barrier (layers of newsprint or cardboard), followed by a 4- to 6-inch layer of organic matter. It's best to start the mulching process early in the spring before the grass has

been mowed. If you need to adjust the pH, sprinkle limestone or sulfur on top of the sod; then mulch with layers of newspaper, compost, and straw. If you can't get straw, use autumn leaves or another organic mulch such as rice or buckwheat hulls. The garden will be ready for planting the following spring. To keep it weed-free, continue to mulch after the plants are in place.

Double Digging

Pioneered in market gardens outside Paris where space is at a premium, this method aerates the soil much deeper than tilling, giving your plants 18 to 20 inches in which to spread their roots. Because they grow downward, plants can be closer together. In addition, their deep roots make them less vulnerable to drought. Double digging is ideal for raised beds.

Spread a few inches of cured compost on the soil. Dig a trench about 10 in. deep, placing the soil in a wheel- barrow. With a garden fork, break up the bottom of the trench as deeply as you can.

Dig a second trench alongside the first, placing the soil you remove in the first trench. Fork the bottom of the trench. Work your way across the bed in this manner until you reach the other side. Dig and fork the last trench, and fill it with the soil from the wheelbarrow.

Improving Your Soil

The ideal soil has a loose, crumbly texture. It holds both air and water well, yet it drains quickly. It should be loaded with numerous kinds of microorganisms and soil-enhancing creatures and an adequate nutrient supply. Finally, it should have an acid/alkaline (pH) balance appropriate for the particular plants being grown.

Because few garden soils start out fitting this description, you'll have to learn how to improve what you have. Most people need to work on their soil's structure, its nutritional qualities, the pH, and drainage. But on a small farmstead scale, it doesn't take long to improve poor soil.

The texture of the soil influences the availability of air, water, and nutrients to all soil life, including plant roots. For example, a very sandy soil drains quite quickly, while a soil containing a large percentage of clay will retain too much water. Although gypsum can be added to clay soil to enhance drainage, you can't fundamentally change the texture of your soil. But you can change its structure.

Structure refers to the way that a soil's particles group together. In a soil with good crumb structure, for example, a random handful looks like a slice of perfect chocolate cake. You can see numerous air spaces between clumps of glistening particles.

Many of these clumps, or soil aggregates, are literally glued together. As soil animals and microorganisms live and die, they excrete sticky substances that bind soil particles together. The channels created by earthworms stay in place for some time, giving air and water a good pathway. Plant roots do this, too. After the plant dies, its roots decompose leaving open channels behind. Both air and water occupy these spaces. If the pore spaces are large, water and air easily move through the soil, ensuring good drainage. The spaces also improve the capillary action of pulling water from lower soil depths in times of drought. As soil organisms break down organic matter, they produce humus, beneficial material that contains decomposed organic matter as well as microbial organisms and their by-products. In nature, humus takes years to form. You can produce it in your garden in a single season. All it takes is the addition of compost.

A Hands-On Soil Test

This simple test gives you a quick take on the structure of your soil. Scoop up some soil, and rub it back and forth in your hand. Then give it a squeeze.

SANDY SOIL is loose and gritty. When squeezed, it falls apart even if moist. Sandy soil is good for potatoes and wet areas but holds moisture poorly in dry periods.

LOAM soil feels smooth, even silky, when wet. When squeezed, it holds its shape somewhat. Loam is ideal for a garden, easy to work, and holds moisture well.

CLAY feels slick, with little grit. When squeezed, it holds its shape. Clay drains poorly and can be difficult to work.

Compost Recipe

Begin with about 6 inches of dry plant material, such as old cornstalks or shredded autumn leaves. Then add a 2- to 4-inch layer of moist green material. Sprinkle on a half-inch layer of garden soil before you add another layer of dry material. If you have manure, add it with the green layer—manure adds nitrogen and microorganisms. Another way to supply microbes is to substitute compost from a previous batch for the garden soil. Or sprinkle a canister of compost-activator between the green and dry layers.

This proportion of dry to green materials gives the decomposing microorganisms the optimal blend of carbon (dry stuff) to nitrogen (green stuff), roughly 25 to 40 times as much carbon as nitrogen. Make the pile at least 3 feet tall. If the materials are very dry, sprinkle them with water until they are about the consistency of a wrung-out sponge. As the materials begin to decompose, the interior of the pile will get hot as the microorganisms do their work. Fold the outside of the pile into the middle. That way, disease-causing fungi and bacteria, as well as most weed seeds, will be exposed to temperatures hot enough to kill them. Compost is considered "finished" when it no longer heats up after it is turned and most materials are decomposed.

Cure the Compost

Unfinished compost can introduce poisons to the soil and deplete nitrogen as microorganisms complete their work. To cure compost, pile it in an out-of-way spot; cover it with a tarp to keep the rain or snow off; and let it sit undisturbed for one to three months. If you plan to use it in a vegetable garden in the spring, you can also cure the compost by spreading it over the bare soil in late fall after the crops are finished; the compost will cure in place over the winter and be ready for spring planting.

Compost "tea" is useful as a fertilizer and treatment for fungal diseases in plants. Make it by steeping compost in water overnight.

Build Your Own Three-Bin Composter

Three bins placed side by side allow you to turn compost into a new area as it decomposes. To hold the front boards in place, cut slots into a 1x4 where the top boards can fit.

Add a top to the bin to keep rainwater out. Make it from plywood, and use 1-inch pipe clamps to hold electrical conduit in place. Drill 1-inch holes in supporting boards for the conduit.

Tools and Materials:
Saw, mallet, level, drill, hammer, 1x6 boards, 2x2 wooden stakes, 1-inch. pipe-clamps, ¾-inch electrical conduit, screws

Optional Hinged Cover

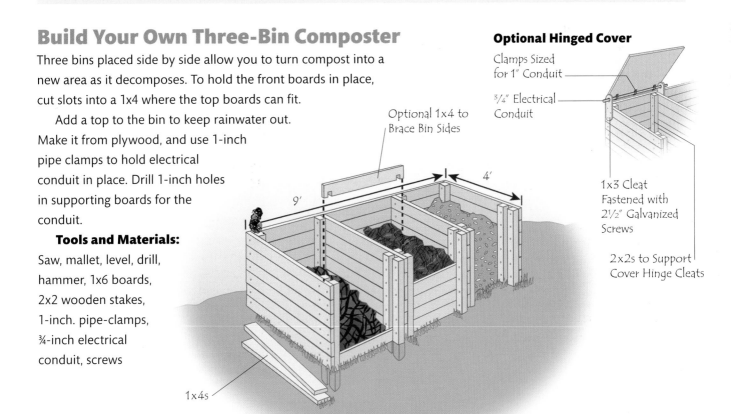

Clamps Sized for 1" Conduit

¾" Electrical Conduit

Optional 1x4 to Brace Bin Sides

1x3 Cleat Fastened with 2½" Galvanized Screws

2x2s to Support Cover Hinge Cleats

9'

4'

1x4s

Testing Your Soil

Correcting the soil pH is the first step toward creating fertile soil. Testing pH measures the acid/alkaline balance in the soil. Lower numbers on the pH scale indicate greater acidity; pH numbers above 7 indicate alkaline soil. Areas with high rainfall tend to have acid soils. Dry regions tend to be alkaline.

The pH level is important because it determines which nutrients are available to plants. In very acid conditions, manganese can become so accessible that it can cause problems. In alkaline soils, plants usually suffer from severe micronutrient deficiencies that can stunt or kill them. A home soil test will give you a rough assessment, but by using your Cooperative Extension Service or a private lab, you'll get more specific results and recommendations for what corrective actions to take.

Do the testing in mid- to late-summer to determine what is available in the soil at the peak of the growing season. Test the soil every year for the first two years of the garden, every other year for the next four to five years. If you've balanced out your soil successfully, you need only test every third or fourth year thereafter.

When you receive your soil test results, you'll learn how to adjust your soil's pH. If the test results show a pH of 6.2 to 7.5, you're in good shape. If the test results

A do-it-yourself pH testing kit will give you a quick idea of whether your soil is acid or alkaline. Test several locations in your garden; you'll find variations from place to place. Amend the soil accordingly.

indicate a pH above 7.5, you will need to bring it down by adding elemental sulfur. If your test results show a pH lower than 6.2, the lab will suggest a particular quantity of ground limestone.

Lime in the Fall

The best time to apply lime is in the fall before the ground freezes. Apply it over a cover crop or before adding mulch. It will break down over the winter. By spring, the soil should be alkaline ("sweet") enough so that fertilizers will be effective. If you must wait to apply lime in spring, try to do it at least two weeks before planting so that it leaches down in time to do some good. Don't spread lime and fertilizer at the same time, as you'll cause some of the nitrogen in the fertilizer to evaporate.

Planting Your Vegetable Plots

Vegetable plots were once planted the way farm fields were—in single rows spaced far apart from each other. This system is useful for crops planted in hills, such as potatoes and corn, but many gardeners and small farmers have abandoned it in favor of more practical beds. Planting vegetables in beds offers these advantages:

- **Compost and soil amendments can be concentrated where plants are growing, rather than being wasted on the pathways.**
- **Building up the height of the bed increases aeration, drainage, and spring heating.**
- **Plants can be spaced so that they shade out weeds, decreasing weeding time.**
- **Plants grow more quickly and vigorously because their soil has not been compacted by being trod upon.**

A bed 4 feet wide allows you to grow four or more rows of a small crop such as lettuce, three rows of a larger crop such as spinach or peas, two rows of big crops such as corn or tomatoes, and a single row for members of the space-hogging squash family.

Seeding Technique

Whether you plant in single rows or rows in a bed, begin by marking the row locations. Stretch string between two stakes as a guideline for planting. This may seem fussy, but a straight row saves time and trouble by making weeding easier and ensures that plants have adequate growing space. Make shallow furrows just under the strings—no deeper than three times the width of the seed.

Space seeds carefully. Most seed packets list a spacing distance, but in many cases this recommendation will require that you thin. Reliable spacing distances are given for each vegetable beginning on page 51. If you follow these suggestions, you usually won't have to bother with the chore of thinning.

Planting Methods

Row planting makes weeding quick and efficient. You can simply hoe down the bed between the rows to uproot the weeds. You will have to do some hand-weeding in the row, of course, but the major part of the job can be done in only a few minutes. However, rows are not the only way to plant beds.

Broadcast seeding is appropriate for small crops like carrots and baby lettuce. Plant by holding the seed loosely in your hand and releasing small amounts as you flick your wrist over the area. After the plants are up and have their second set of leaves, thin them to the recommended in-row spacing. Broadcast planting offers plant density and yields per square foot much greater than with a row system. But to work well, the soil must be largely free of weeds. If it isn't, you'll have to do too much handweeding—time-consuming and no fun. Carrots, beets, scallions, radishes, salad greens, baby turnips, and spinach broadcast well.

Among its many advantages, a raised bed is easy to access without stepping on and compacting the soil.

43

Handling Small Seeds. It's easy to handle large seeds such as beets, peas, and corn. Even the smaller broccoli and cabbage seeds won't cause you problems. But tiny seeds, such as carrot, can be hard to space well. If you have difficulty getting a sparse enough seeding, mix the carrot seed with equal portions of dry sand. Take up pinches of this mixture, and sprinkle them along the furrow. The tiny seed should be spaced well with this method.

Seeds must be consistently moist to germinate. In the spring, when the soil is wet and rains are frequent, this is rarely a problem. Water just after you plant, using a fine misting nozzle. Check the soil each morning and evening until the seeds germinate to see whether you need to water again. But if you're planting seeds in hot weather or in a drought, you'll have to cover the soil with burlap or newspaper to hold moisture. (See "Keeping Seeds Moist," on the opposite page.)

Seeds that require light to germinate often dry out if you leave them on the soil surface. Avoid this problem by filling your furrow with vermiculite. Plant the seeds on the surface of the vermiculite, and then water with a misting nozzle. The water will work the seeds into niches in the vermiculite so that they remain wet, but enough light will penetrate the vermiculite so that they germinate.

Straw-Bale Gardening

If you have some hard, rocky soil or an area of old patio you'd like to make productive, here's a quick solution. Choose a location and measure the area you'd like to convert to a garden. Bales vary in size, but those that are 18 inches wide by 36 inches long by 14 inches high are common, and at about 50 pounds, they are manageable. Decide the number you'll need and be sure to buy straw, not hay, bales. (See page 154.) Try to find bales with synthetic twine so they will last for several seasons without falling apart. Arrange them so that you can easily reach to the center of your straw-bale bed. Remember, the bales are not an enclosure. You'll be planting directly on top of them.

If you set the bales on grass or weeds, lay down 6 or 8 layers of newspaper or a layer of landscaping fabric first. Place the bales cut size up—the better to pound in stakes and get the water to penetrate. Soak them with water repeatedly for several days. They'll heat up initially. Wait for them to cool; then top dress with about 3 inches of potting soil.

Plant lettuce, herbs, tomatoes, cucumbers, peas, squash—most anything but deep-root crops. Sets can be planted by digging a small hole in the potting soil.

You can also make a cold frame from bales by positioning them in a U shape to shield young plants from the wind. Add sheet plastic to make a mini greenhouse, but be sure to remove it on sunny days to avoid cooking the plants. Bales also make a handy compost pile. Inner sides of the bales can be sloughed off as they rot into the compost through the winter, leaving a handy pile ready for applying to the garden in the spring.

Transplanting

Many vegetable seedlings are transplanted directly into the garden. Place vegetables set into the soil at the same depth they grew in their starting containers. Tomatoes are one exception. By burying a few inches of the stem, roots will grow from the stem, further stabilizing the plant and increasing its ability to obtain food and water.

Vegetable seedlings require immediately available supplies of phosphorous, especially in cold soil. Soak the flats or root-balls of your transplants in a solution of liquid seaweed and fish emulsion, diluted as recommended on the bottle, for at least 15 minutes before planting. Later in the season when the soil is warm, this root drench will keep the root-ball wet while the plant is adjusting to its new environment.

To avoid damaging tender stems from leaning them sideways, carry them upside down, supporting the plant as you set it in the ground.

Keeping Seeds Moist

In drought conditions, consider planting in deep furrows. Dig a deeper-than-normal furrow, and water it well. Then plant in the bottom as usual. Cover the seed with a thin layer of soil, and water again, using a misting nozzle to avoid displacing the soil. Or plant in a normal furrow, and cover them with row-cover material or untreated burlap. Saturate the covering with water several times a day. If the seeds germinate in the dark, you can use a layer of wet newspapers, but check for germination every morning and evening.

Extending the Seasons

Season extenders such as row covers, cages, cold frames, and glass cloches help you grow plants before and after the normal growing season. They also give some protection from sudden cold snaps. Season extenders vary tremendously in effectiveness and cost. Some are most useful for covering large areas when a light frost threatens, while others can actually protect plants from temperatures in the low 20s.

Whether you use traditional glass cloches or cut off the bottom of a plastic bottle, the principal is the same: extending the growing season by sheltering your plants from late frosts.

Row covers are among the most handy. They can be put up quickly and, because they "breathe," do not have to be removed until bright and hot weather sets in. Use two layers of row-cover material whenever temperatures threaten to fall below 28°F. You'll be getting at least 6°F and possibly as much as 8°F of protection, meaning that your plants will not only survive, they will also suffer less than those that just make it through the low temperatures.

Know Your Zone

The best place to start crop planning is with the USDA Hardiness-Zone Map and the American Horticultural Society's Heat-Zone Map. The Hardiness-Zone Map is based on records of the average minimum temperatures all over the U.S. and Canada. If you learned what your zone was long ago, you might want to check it again. The boundaries of many of the zones were changed in 1990, based on actual changes in the climate since 1965. The "A" and "B" divisions in each zone are a further refinement. For example, even though both the Berkshire Mountain area and the Pioneer Valley of Massachusetts are listed as Zone 5, the mountains are 5A while most

of the valley is 5B. The higher altitude makes mountain winters significantly colder than they are in the valley, only a few miles away—a microclimate in action.

The Hardiness-Zone Map is invaluable when you are shopping for seeds and plants. All perennials are rated according to the zone to which they are winter hardy. In general, you are safe if you stick with plants having the same rating as your zone. But you'd be wise to make an exception to this rule when choosing very expensive stock and fruit trees that bloom early in the season. In these cases, choose plants that are rated for the next colder zone. This is when the "A" and "B" ratings come in handy, too.

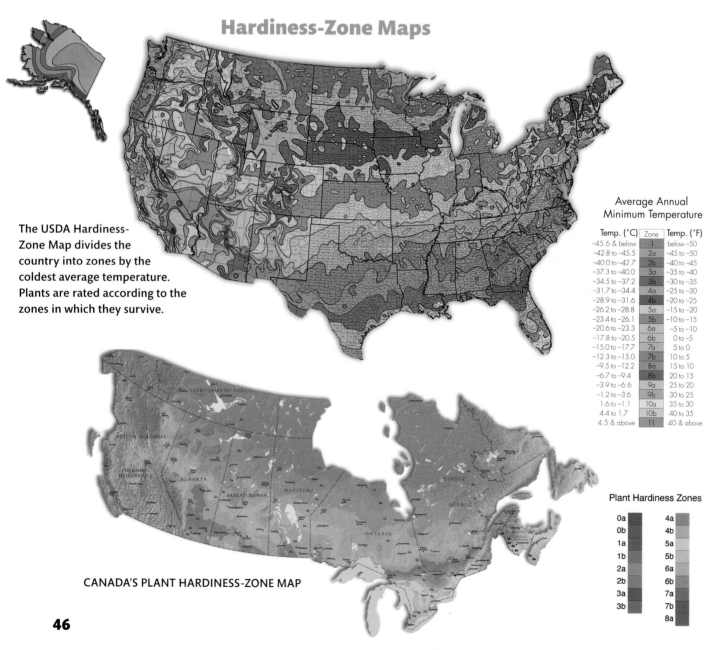

Hardiness-Zone Maps

The USDA Hardiness-Zone Map divides the country into zones by the coldest average temperature. Plants are rated according to the zones in which they survive.

Average Annual Minimum Temperature

Temp. (°C)	Zone	Temp. (°F)
−45.6 & below	1	below −50
−42.8 to −45.5	2a	−45 to −50
−40.0 to −42.7	2b	−40 to −45
−37.3 to −40.0	3a	−35 to −40
−34.5 to −37.2	3b	−30 to −35
−31.7 to −34.4	4a	−25 to −30
−28.9 to −31.6	4b	−20 to −25
−26.2 to −28.8	5a	−15 to −20
−23.4 to −26.1	5b	−10 to −15
−20.6 to −23.3	6a	−5 to −10
−17.8 to −20.5	6b	0 to −5
−15.0 to −17.7	7a	5 to 0
−12.3 to −15.0	7b	10 to 5
−9.5 to −12.2	8a	15 to 10
−6.7 to −9.4	8b	20 to 15
−3.9 to −6.6	9a	25 to 20
−1.2 to −3.6	9b	30 to 25
1.6 to −1.1	10a	35 to 30
4.4 to 1.7	10b	40 to 35
4.5 & above	11	40 & above

CANADA'S PLANT HARDINESS-ZONE MAP

Plant Hardiness Zones

0a	4a
0b	4b
1a	5a
1b	5b
2a	6a
2b	6b
3a	7a
3b	7b
	8a

Heat-Zone Map

Researchers have recently discovered that plants begin to suffer cellular damage at temperatures over 86°F. The American Horticultural Society's Heat-Zone Map, introduced in 1998, divides the United States into 12 zones based on the average number of "heat days," or days over 86°F that a region experiences. The zones range from Zone 1 (no heat days) to Zone 12 (210 heat days a year). More and more plant breeders are now labeling plants with both USDA Hardiness Zones and American Horticultural Society Heat Zones. Just as various factors affect a plant's ability to survive cold, so are there factors that affect a plant's ability to survive heat, including humidity. Use both maps as starting points. Then allow your own experience and the experience of other gardeners in your area to guide you in your choice of plants.

Learn Your Frost-Free Dates

Frost damage can be a threat to plants at both ends of the season. In the fall, early frosts end the growing season for many of the fruiting vegetable crops.

Tomato, pepper, eggplant, snap bean, summer squash, cucumber, and melon plants all blacken and die if their leaf temperature falls below 32°F. But often, the first autumn frost is only a teaser. After that one clear, cold night, you may have several more weeks of good growing weather. As a first step, you can position your less hardy vegetable plants so that frost is less likely to damage them. If possible, place them near the crest to halfway down a south-facing slope. See page 45 for how to protect your crops from light frosts with season-extending products and techniques.

By the same token, in the spring, late frosts often kill the blossoms of fruiting plants. Peaches, apricots, nectarines, and early strawberries are the most vulnerable, but apples, pears, and plums can also be harmed by an extremely late frost. Here again, location matters. If your fruit trees are positioned on a slope, it's likely that a light frost will roll right over your garden, doing only minimal damage. However, if the fruit trees are planted at the bottom of a hill, the frost may settle on your plants, causing serious problems.

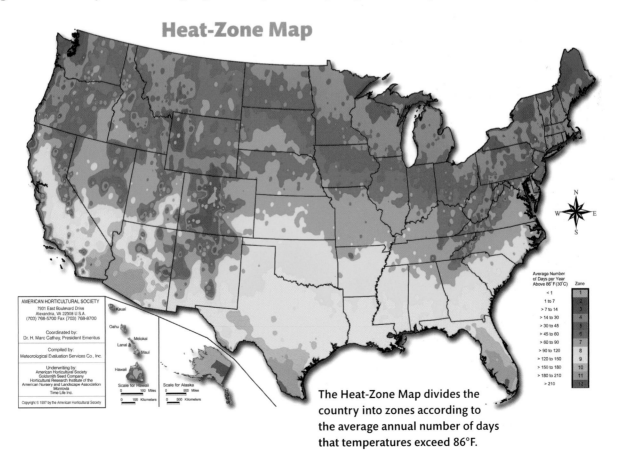

Heat-Zone Map

Average Number of Days per Year Above 86°F (30°C)	Zone
< 1	1
1 to 7	2
> 7 to 14	3
> 14 to 30	4
> 30 to 45	5
> 45 to 60	6
> 60 to 90	7
> 90 to 120	8
> 120 to 150	9
> 150 to 180	10
> 180 to 210	11
> 210	12

AMERICAN HORTICULTURAL SOCIETY
7931 East Boulevard Drive
Alexandria, VA 22308 U.S.A.
(703) 768-5700 Fax (703) 768-8700

Coordinated by:
Dr. H. Marc Cathey, President Emeritus

Compiled by:
Meteorological Evaluation Services Co., Inc.

Underwriting by:
American Horticultural Society
Goldsmith Seed Company
Horticultural Research Institute of the
American Nursery and Landscape Association
Monrovia
Time Life Inc.

Copyright © 1997 by the American Horticultural Society

Kauai
Oahu
Lanai Molokai
       Maui
Hawaii

Scale for Hawaii
0   100 Miles
0   100 Kilometers

Scale for Alaska
0   500 Miles
0   500 Kilometers

The Heat-Zone Map divides the country into zones according to the average annual number of days that temperatures exceed 86°F.

Saving Seed

Until the 1940s, all gardeners and farmers saved their own seeds. The seeds were open-pollinated varieties, meaning their parents were genetically similar, but not identical, to each other. The plants that grew from these seeds were "true to type" because they carried the characteristic traits of their parents. These plants are called heirlooms.

Hybrid seeds result from crossing two or more plants of the same species that are genetically different. A hybrid seed produces a plant that is different from each of its parents but almost identical to every other plant grown from the same cross.

Ironically, open-pollinated plants of the same variety usually have more genetic diversity between individual plants than do hybrids. But hybrids make life much more convenient—such as knowing that all the broccoli will mature within a few days of each other, for example.

Heirloom tomatoes, top right, and eggplants, center right, offer fresh colors, flavors, and textures—and keep an important heritage of self-sufficiency alive. Heirloom beans like these cranberry pole beans, bottom right, are a reminder of how important this long-storing staple was for nineteenth-century farmers.

Benefits of Hybrids versus Open-Pollinated Plants

Most of the seeds you find in catalogs are hybrids that have been bred for certain characteristics. In some cases, these plants are superior to any open-pollinated crop you can grow, but in others, texture and flavor have been lost. Some hybrids used by commercial farmers have another limitation—seeds are being bred that will not grow unless they are treated with specific fertilizers and pesticides.

As the gene pool shrinks, not only are thousands of wonderful varieties being lost in favor of the hybrids that are replacing them, the genetic similarity of many crops is making them more vulnerable to catastrophe. In 1970, for example, more than half of the U.S. corn crop in the South was stricken with

a fungal disease; overall, more than 1 billion bushels were lost. All of the affected corn shared parentage. This disaster, as well as the loss of 40 percent of the Soviet Union wheat crop in 1971, which was also due to genetic uniformity, alerted farmers and gardeners alike to these serious problems.

In response, people began to save seeds, just as their grandparents and great-grandparents once did. Seed-saving groups are now selling open-pollinated seeds as well as trading them between members. The Seed Savers Exchange, **www.seedsavers.org,** based in Decorah, Iowa, has a wealth of information about saving your own seed and offers a wide range of open-pollinated seeds for sale.

Preventing Weeds, Pests, Diseases

Weeds can be the most troublesome crop problem you'll experience, particularly during the first few years of your garden. You've no doubt heard the cliché about weeds being nothing more than plants "out of place." But no matter what your relationship to the weeds in your garden, you'll be able to control and, believe it or not, use them better if you understand them.

Weeds are tenacious because they have remarkable reproductive capabilities. Most weeds produce amazing numbers of seeds. For example, redroot pigweed can drop more than 100,000 seeds in a square-foot area, and common mullein routinely produces more than 200,000 seeds. These seeds have a long period of viability, some able to germinate for up to 50 years. Vines such as poison ivy and morning glory can send up new shoots from rhizomes that have traveled 20 feet from the mother plant, perhaps under a sidewalk or a concrete-block wall. Plants such as quack grass and curly dock can reproduce from a tiny piece of root as readily as from seeds.

Put Weeds to Work

Use the fast growth of weeds such as lamb's-quarters to provide a quick green manure crop. Or take advantage of weeds with strong, deep roots like amaranth to break up hard pans, increase water movement through the soil, and expand the area that crop roots can penetrate. As long as you mow or till before the weeds set seed, you can benefit from this "free" crop.

Keep ahead of the weeds that sprout up in your garden beds to give your vegetables the best chance to thrive.

Weed Defense

In comparison with weeds, most of our cultivated plants are weak creatures. When placed side by side, you can count on the weeds to take more than their fair share of space, above- and below-ground. Here are some ways to defend your garden from weeds:

Hoe or cultivate as shallowly as possible. Weed seeds buried deep in the soil will germinate if you bring them to the surface.

- **Weed early and often. It's easier to remove tiny seedlings than foot-tall monsters.**
- **Mow or deadhead weeds or cultivated plants, such as dill or mint, which can become weeds.**
- **Remove most weeds from the garden after you pull or hoe them.**
- **Never till a patch of perennial weeds that can reproduce from the rootstock. Instead, dig out the roots or mow repeatedly.**
- **Add mulch. It blocks the sunlight necessary for weed seeds to germinate. Some mulches such as sawdust or wood chips rob so much nitrogen from the soil as they decompose that weeds can't survive. Low-growing companion plants that you seed between your ornamental plants or crop rows can also keep weeds down.**
- **Take advantage of natural plant antagonisms with companion plantings. Winter rye is often** used to inhibit the growth of quack grass. Asters slow down ragweed, and barley hinders redroot pigweed, purslane, and ragweed.
- **Plant intensively. If plants are spaced so they don't compete with each other but still shade the soil surface below them, weeds can't get a foothold once the cultivated plants mature.**
- **Rotate your crops. In an area where weeds have built up over the season, plant a heavily mulched crop the following year. Similarly, you can take advantage of beds that were mulched heavily by planting a crop that doesn't compete well against weeds, such as carrots.**

Preventing Pests and Diseases

Prevention is the secret to having a productive garden, free of pests and diseases. For specific problems, consult with your County Extension Service, but here are some general preventative practices:

- Encourage a diverse and complex group of living creatures, from insects and microorganisms to plants, birds, and frogs.
- Build and maintain soil health by using composts, organic mulches, cover crops, green manures, and natural soil amendments such as rock powders in an appropriate manner.
- Avoid using materials such as herbicides, synthetic fertilizers, or insecticides that kill or inhibit beneficial organisms.
- Avoid plant diseases by rotating crops, choosing disease-resistant cultivars, and providing the correct conditions for each plant you grow.
- Provide the best possible environment for every plant you grow, from soil characteristics to air circulation and drainage.
- Design your garden plots to be as easy to care for as possible. Plant only as much as you can easily maintain.

Vegetables, Plant by Plant

Here is the essential information you'll need for growing vegetable crops. Check the variety you choose for specifics on spacing and maturation. See pages 46–47 for the USDA Hardiness-Zone Map and the American Horticultural Society's Heat-Zone Map. Both are helpful in choosing the variety right for your locale.

ONIONS

Onions with the tops knocking and shriveled are ready for storage.

GARLIC

LEEKS

ALLIUMS: GARLIC, LEEK, ONION, SCALLION

TEMPERATURES

Moderate. **Onion** bulbs may be watery if temperatures rise above 85°F when they are forming. **Garlic** cloves must be chilled to freezing or below to form bulbs. **Leeks** tolerate frosts when mature and can over-winter with protection.

SOIL

Onion family vegetables like moist, humus-rich, fertile, well-drained soil with a pH of 6.0 to 7.0.

PROPAGATION

Onions. Plant seeds indoors, 8 to 10 weeks before the frost-free date and transplanted to the garden a week or so before that time. Keep soil temperatures at 70° to 75°F during germination. In Zones 5 and warmer, direct-seeded onions will have time to form bulbs. Sets can also be used, but the resulting onions will not keep well through the winter.

Leeks and **Scallions.** Plant from seed, either early indoors or directly in the garden.

Garlic. Plant cloves in autumn when you plant tulips—early October in Zones 3 and 4, mid-October in Zones 4 to 5, and late October in Zones 7, 8, and 9. Plant several inches below the soil surface, and mulch with 6 to 12 inches of straw or rotted leaves once the top surface of the soil has frozen. To plant in spring, prechill the cloves for a month in the refrigerator, and plant as soon as you can dig a hole in the ground.

SPACING

Onions, 6 to 8 inches between large Spanish onions and 4 to 6 inches between storage onions. **Scallions,** ½ inch apart in double rows 1 foot apart. **Garlic,** 6 to 8 inches in rows at least 1 foot apart. **Shallots,** 6 inches apart in rows 1 foot apart.

CULTURE

In soils with moderate fertility, drench rows of all alliums with compost tea or a seaweed/fish emulsion once a month until late summer. Cultivate shallowly because roots, particularly those of leeks, are near the surface.

Onions and **Leeks.** Trim leaves if they threaten to fall over in their starting containers and again when you transplant them to the garden, cutting so top growth is no more than 6 inches tall. Hill around leeks to increase the length of the white stem. To avoid damaging the roots, take the soil for hilling from the walkways or another section of the garden rather than from between the rows.

Garlic and **Shallots.** In early spring, pull back the mulch to see whether shoots are appearing, but leave the mulch along the row so that you can re-cover the plants if frost threatens. Plants may die if the leaves are heavily frosted. Similarly, they rot easily if too wet, so pull mulches back whenever possible.

AVERAGE YIELDS

Onions, 1 pound per row foot.
Scallions, ½ pound per row foot.
Leeks, 1½ pounds per row foot.
Garlic, ¼ to ½ pound per row foot.

HARVEST

Onions. For fresh use, harvest onions as soon as they are usable. For storage onions, check to see that all the tops have fallen over in late summer, and knock over any that haven't. Wait until the tops dry and the bulbs have formed a tight neck. Then pull and place in a single layer in a warm, dry area with good air circulation. Wait until the skins rustle before storing.

Scallions. Pull scallions as soon as they reach a harvestable size.

Garlic. Harvest fall-planted garlic bulbs in mid-July in most regions. Count leaves to be certain that there are (or have been) at least seven, meaning that the bulb is covered with seven layers of skin. If harvesting for seed, wait until the end of the month to harvest. Dig the bulbs only after several days of dry weather. Wait until fall to dig spring-planted garlic.

Shallots. Handle fall-planted shallot sets as you do fall-planted garlic and direct-seeded crops as you do spring-planted garlic.

ASPARAGUS

TEMPERATURES

Versatile. Asparagus grows in Zones 2 to 9.

SOIL

Asparagus like deep, well-drained, fertile soil that is rich in humus with a pH of 6.5 to 6.8. Add 2 inches of fully finished compost to the growing area before planting.

PLANTING

Buy 1-year-old crowns or start seeds outdoors as soon as the soil temperature is 50°F and evening lows are in the high 30s or low 40s. Plants naturally enlarge from the crown each year, giving you more spears.

SPACING

Space crowns 2 inches apart in rows 4 to 6 inches apart.

CULTURE

Soak roots in compost tea for several hours before planting. Plant in early spring, as soon as the soil temperature is 50°F. If your soil is soggy, heavy, or compacted, make raised beds. Dig a trench, about 8 to 10 inches deep, and make mounds of compost-enriched soil every 2 feet along its length. Drape the roots of the crowns over the mound, spreading them like octopus arms. Cover them with 2 to 4 inches of soil. Every two to three weeks during the first season, add another couple of inches of soil to the trench until it is slightly raised above the surrounding soil surface. Let the ferns grow undisturbed the first year. Keep asparagus beds well weeded. Every fall, cut the dried ferns off at the soil surface, and heap 2 inches of compost over the bed. Mulch deeply

ASPARAGUS

as soon as the ground freezes. In spring, pull back the mulch when spears begin to appear.

AVERAGE YIELDS

Expect 3 to 4 pounds per 10-foot row. Plant at least 20 feet per person in the household.

HARVEST

Do not harvest the first year. Harvest lightly during the second year. During the fourth year, you can count on a harvest season of three to four weeks and eventually build to six to eight weeks. To harvest, snap off the spears at or just below ground level when they are 6 to 10 inches high and the tips are still closed. Best storage results when spears are harvested in very early morning.

Alliums: Garlic, Leek, Onion, Scallion; Asparagus; Bean: Pole Bean, Lima Bean, Soy Bean

POLE BEANS

LIMA BEANS

SOY BEANS

BEAN: POLE BEAN, LIMA BEAN, SOY BEAN

TEMPERATURES

Warm. Beans prefer minimum night temperatures of 40°F and minimum day temperatures of 50°F, but plants perform better in warmer conditions. Frost kills both seedlings and mature plants.

SOIL

Beans require well-drained, good humus content, moderately fertile soil with a pH of 5.5 to 6.8.

PLANTING

Plant seeds in late spring or early summer, after danger of frost has passed and soil temperature is 60°F. For a continuous harvest, plant short-season cultivars every two to three weeks until midseason. Yields will be higher if you dust the seeds with a "bean and pea" inoculant powder (available at garden centers) before planting.

SPACING

Plant seeds 4 to 6 inches apart in rows 1½ to 2½ feet apart. In 4-foot-wide beds, plant two rows.

CULTURE

Bacteria on the roots fix nitrogen for beans, so they can thrive in low-nitrogen soils. However, they do require high levels of potassium and phosphorus, supplied by greensand and rock phosphate, respectively. Test and adjust soils if necessary. Keep soil moist but not soggy. If diseases have been a problem, water only the soil, not the foliage. Never walk near the beans when foliage is wet. In areas where pests have been troublesome, cover plants with row-cover material until they are ready to harvest.

AVERAGE YIELDS

Bush beans yield about 8 pounds per 10-foot row; **pole beans** yield about 15 pounds per 10-foot row; and dry beans yield about 8 pounds per 100-foot row.

HARVEST

Keep beans picked to prolong flowering and fruiting. Pick beans with the stem attached to lengthen storage time.

BEET

BEET

TEMPERATURES

Beets like moderate temperatures but prefer nights of 55° to 60°F and days of 65° to 75°F—will tolerate temperatures of 55° to 85°F. Mulch in warmer temperatures.

SOIL

Beets like well-drained, moist, highly fertile, and friable soil with a pH of 6.0 to 6.8. Beets do not tolerate boron deficiencies. If soil is slightly deficient, spray plants with liquid seaweed every two weeks.

PLANTING

Direct seed when soil temperatures are at least 45°F and air temperatures are 50°F or above. For successive harvests, plant every three weeks until midsummer. Plant storage beets about 60 to 70 days before your first expected fall frost.

SPACING

Plant seeds 3 to 4 inches apart in rows 1 foot apart. Because several seeds are enclosed in each "seed," you will need to thin seedlings to stand 4 to 6 inches apart in the row. Wait to thin until the thinnings can be used in salads or cooked as baby greens.

CULTURE

Beets require steady supplies of moisture and nutrients. In lean soils, water them with a compost tea or a solution of liquid seaweed/fish emulsion every two weeks.

AVERAGE YIELDS

For moderately sized beets, expect about 1 pound for every row foot.

HARVEST

Thin plants successively to harvest baby beet greens, then baby beets, and finally mature beets. Cut tops from beets, about 2 inches above the beet itself, as soon as you harvest, and store separately. Beets that will go straight to the table can be allowed to stand in moderate frosts. However, for the best-quality storage roots, pull them several days before your first expected frost.

BROCCOLI

BRUSSELS SPROUTS

CABBAGE

KALE

BRASSICAS: BROCCOLI, BRUSSELS SPROUT, CABBAGE, CAULIFLOWER, KALE

TEMPERATURES

Cool. They like 50° to 60°F nights, 60° to 75°F days. **Broccoli** may "button," or form small heads at lower temperatures or form small, loose heads prematurely at warmer temperatures.

Cauliflower is more temperature-sensitive than other members of the family. Germinate seeds at a soil temperature of 70° to 75°F, and keep seedlings at night temperatures of 50°F and day temperatures of 60°F.

Kale and **Brussels sprouts** tolerate freezing temperatures. Use row covers to harvest kale until midwinter.

SOIL

Provide nutrient- and humus-rich, well drained soil that is high in calcium, with a pH between 6.7 to 7.2.

PLANTING

For first crop of **broccoli, cauliflower,** and **cabbage,** start seeds indoors about two months before last spring frost, keeping them in their ideal temperature range.

Harden off and transplant 5- to 6-week-old seedlings about two to three weeks before last frost under row-cover material. Plant fall crops so that they will mature just at the time of the first expected frost. They require about two weeks longer to mature in the fall but keep their quality in the field much longer than they do in the spring.

Brussels sprouts. Plant or transplant Brussels sprouts only once a season, timing it to mature just at the first expected frost date.

Beet; Brassicas: Broccoli, Brussels Sprouts, Cabbage, Cauliflower, Kale

Kale. Direct-seed kale four to six weeks before last spring frost. Plant fall crops two months before the first expected fall frost. To transplant, start in peat pots or soil blocks six weeks before the frost-free date.

SPACING

Plant **broccoli, Brussels sprouts, cauliflower,** and **cabbage** 1½ to 2 feet apart in rows 1½ to 2 feet apart.

Kale. Plant kale seeds about 3 to 4 inches apart. Thin plants to stand 12 to 15 inches apart.

CULTURE

Many pest problems can be avoided by covering newly transplanted or seeded crops with row-cover material and leaving it in place as long as temperatures are 75°F or below. Keep soil evenly moist and weeds under control. In moderately fertile soils, drench the soil with seaweed/fish emulsion every three weeks; apply about 1 cup per row foot or foliar feed once every three weeks with a seaweed/fish emulsion early in the morning.

Brussels sprouts. When the sprouts begin to develop, snap off the leaf stalks just under the sprouts that have begun to grow. Do this every two weeks throughout the season. The plant will look like a strange sort of palm tree by late summer, with knobs all along the stem and a tuft of leaves at the top. About two weeks before frost, pinch off the central growing tip of the plant to encourage it to put energy into maturing the sprouts.

Cauliflower heads turn yellow if the sun hits them. Avoid this by "banding" them. As soon as the developing head is the size of a silver dollar, pull the leaves together over it, and secure them with a rubber band. Peek inside the leaves every couple of days to determine when to cut the heads. Some cultivars are said to be "self-wrapping," meaning that they don't need this treatment. But band them anyway to be on the safe side.

AVERAGE YIELDS

Broccoli, 7½ pounds for every 10 row feet or one main head and 6 to 8 small side shoots per plant. **Brussels sprouts,** ¾ pound for every row foot. **Cauliflower,** 1 head per plant. **Cabbage,** 1 head per plant. **Kale,** ¾ to 1 pound per row foot.

HARVEST

Broccoli. Cut off the main head when it looks fully formed and while buds are still tight. Best quality is ensured by cutting in early morning before plants have warmed. For maximum side-shoot production, cut the main head with a short stalk. Cut side shoots as they develop. In most cultivars, they will continue to develop for up to six weeks beyond the time you cut the main head.

Brussels sprouts. Light frost improves the flavor, but if sprouts are left on the plant too long, they loosen and open. To harvest, push the sprout to one side with your thumb, holding the plant steady with your other hand.

Cauliflower. Cut cauliflower when the heads have fully expanded, while the curds are still tightly bunched together. Even though this may seem difficult to determine, a little experience with various cultivars will make this easy. Cauliflower will not make side shoots.

Cabbage. Cut whole cabbage heads when they feel tightly packed when you gently squeeze them. If you cut above the crown, up to four smaller heads will form.

Kale. Pick kale early in the morning when plants are cool. Pick whole plants while thinning the rows. After that, harvest bottom leaves by pulling down sharply on their petioles.

SPECIAL CONSIDERATIONS

Brussels sprouts. If late-season aphids are a problem with your Brussels sprouts, direct-seed a tall variety of dill between the plants. The dill will feed aphid predators, keeping the pests in check.

Cauliflower is more difficult to grow in the spring than in the fall, primarily because night temperatures can be too low during the early part of their growth and day temperatures can be too high while heads are forming. Minimize problems by choosing early-maturing cultivars meant for spring and covering plants with row-cover material until it is warm. In Zones 5 and northward, certain cultivars can make good heads in the summer.

Cabbage. Choose several cabbage cultivars for variety. If splitting after a rain is a problem, harvest large heads.

Kale. All species of kale have the finest flavor after frost and also stand up to repeated freeze/thaw cycles without losing quality. Where deer are a problem, cover the plants with row-cover material.

SWEET CORN

SWEET CORN

TEMPERATURES

Warm. Corn grows best in night temperatures of 60°F or above and day temperatures of 75°F and above.

SOIL

Corn requires soil that is highly fertile, deep, contains organic matter, and with a pH of 6.0 to 6.8.

PLANTING

Plant where it is to grow, once danger of frost has passed. Seed germinates poorly at soil temperatures below 65°F, but plastic laid over the soil surface for a few weeks or so just before the frost-free date, will raise temperatures sufficiently. (Check the temperature with a soil thermometer before planting.) In Zones 5 and cooler, earlier planting is possible if you plant seeds in a 6-inch furrow, cover them with only an inch of soil, and then cover the furrow with clear plastic. Let seedling plants grow under the plastic, venting it if necessary during the day to prevent fungal diseases. Remove the plastic when temperatures stabilize. In Zones 5 and south, plant a succession of cultivars to prolong harvest.

SPACING

Plant 8 to 12 inches apart in rows 2 feet apart. Plant in blocks of at least 4 rows to ensure proper wind pollination.

CULTURE

Corn is an extremely heavy consumer of nitrogen. If soil nitrogen is low, side-dress with compost every month or drench the soil with 1 cup of compost tea or liquid seaweed/fish emulsion solution per row foot. Make certain soil moisture is steady once plants tassel.

AVERAGE YIELDS

An average of 1 ear per row foot.

HARVEST

When silks have dried and ears feel full, peel back a few husks and puncture a kernel with your thumbnail. A milky-colored liquid should spurt out. Pull ears downward to harvest.

CRUCIFERS: CHINESE CABBAGE, COLLARDS, BOK CHOY, RADISH, RUTABAGA, TURNIP

TEMPERATURES

Cool. Nights of 40° to 50°F and days of 55° to 70°F are ideal although plants will tolerate light frosts and temperatures as high as 80°F.

SOIL

These vegetables require moderate fertility with steady moisture, high humus, and a pH of 5.5 to 6.8. Rutabagas and turnips require balanced amounts of potassium, phosphorus, and boron, and a pH of 6.0 to 7.0.

PLANTING

Chinese cabbage, Bok Choy, and **Collards.** Start indoors four weeks before the frost-free date in peat pots or soil blocks. Direct seed two weeks before the frost-free date. Plant fall crops two months before the first expected frost.

Radish. Plant small numbers of radish seeds where they are to grow, beginning about three to four weeks before the last expected spring frost and continuing each week until about two weeks after the frost-free date. Plant fall crops two months before first expected fall frost, and continue until a week before.

Rutabaga. Plant in mid- to late July.

Turnips. Direct seed turnips three to four weeks before the last expected frost and every week after until temperatures are too warm. For fall crops, plant two months before the first expected frost.

SPACING

Chinese cabbage, collards, and **bok choy.** Plant 6 inches apart and harvest to leave 1 to 1½ feet between plants.

Radish. Plant ½ inch apart in triple rows 1 foot apart. Thin to allow 1 inch per plant.

Rutabaga and **Turnip.** Plant 4 inches apart and harvest to 6 to 8 inches between storage cultivars.

CULTURE

If leaves look yellow, drench the soil with a cup of compost tea per row foot or spray foliage with a seaweed/fish emulsion. Cover plants with row-cover material in the spring to prevent cabbage-root-fly maggot and flea-beetle damage.

Sweet Corn; Crucifers: Cabbage, Collards, Bok Choy, Radish, Rutabaga, Turnip

CHINESE CABBAGE

COLLARDS

BOK CHOY

AVERAGE YIELDS

Per row foot: **Chinese cabbage:** 1 to 3 pounds; **Bok choy, collards,** and **Turnip:** ½ to I pound; **Radish:** 1 to 2 bunches; **Rutabaga:** 1½ pounds.

HARVEST

Chinese Cabbage, Bok Choy, Collards. Cut entire plant at the crown or, for prolonged harvest, pick outside leaves as needed.

Radishes and **Turnips.** Pull before roots are so large that they split. Cool radishes in water immediately after harvesting, and remove the leaves if you plan to store them for more than a day.

Rutabaga. Rutabaga tastes better after a frost. Pull them on a dry day. If you are storing, do not wash, but let the soil dry and brush off excess.

RADISH

RUTABAGA

TURNIP

ZUCCHINI

CUCUMBER

CANTALOUPE

WATERMELON

CUCURBITACEAE: CUCUMBER, GREEN SQUASH (SUMMER SQUASH, OR ZUCCHINI), HYBRID CANTALOUPE, WATERMELON, WINTER SQUASH

TEMPERATURES

Warm. Night temperatures of 55° to 65°F and day temperatures of 75° to 85°F are ideal.

SOIL

Plants require rich fertility, well-drained, moisture-retentive soil with high humus levels and a pH of 6.0 to 6.8.

PLANTING

Cucumbers, Summer Squash, and **Winter Squash.** In Zones 7 southward, start seed where plants are to grow once the soil is 55° to 60°F. In Zones 6 northward, plants may be started indoors in 4-inch plastic or peat pots about three to four weeks before the frost-free date. Place 3 seeds in each pot, and thin to the best two seedlings. Transplant a week after the frost-free date.

Melons. Start melon seeds indoors no more than three weeks before your frost-free date in 4-inch peat or plastic pots. If possible, set seedlings under artificial lights, and give them 16 hours of light per day. Melons may also be direct-seeded once the soil is 60°F.

SPACING

Cucumbers, Summer Squash, and **Winter Squash.** Grow two plants in each hill. Space hills 3 to 4 feet apart in rows 4 to 6 feet apart, depending on size of mature plant.

Melons. Plant two to three melon seedlings or 3 to 5 seeds in hills spaced 6 feet apart in all directions.

CULTURE

Cover plants with row-cover material until they begin to bloom. In Zones 6a and cooler, black paper or plastic mulch will increase yields. Tunnels made of slitted polyethylene or row-cover material also increase yields in the North by increasing temperatures. Remove the tunnels once flowering begins to allow insect pollination. Keep plants consistently moist, but avoid watering foliage to keep diseases down.

AVERAGE YIELD

Cucumbers, about 12 pounds per 10-foot row. **Summer squash,** 2 pounds per row foot at a minimum. **Melons,** about 1 pound per row foot. **Winter squash,** 2 pounds per row foot.

HARVEST

Cucumbers. Pick while they are still slender and completely green; do not let the spot where they rest on the soil surface become yellow or they will be seedy. Cut with the stem for longest storage life.

Melons. Some cantaloupes "slip" from their stems when they are ripe. Test them by gently pushing against the stem every day or so once they look mature. The pale spot on the bottom of other melons, including watermelons, will be distinctly yellow when the melon is ripe and the tendril just above the fruit's stem will have withered and died back.

Summer Squash. Cut from plants, taking at least an inch of stem. Pick daily, taking young fruit that is no longer than 8 inches; 4 to 6 inches in diameter for round cultivars.

Winter Squash. Check for maturity by noting the size, skin toughness, and color, particularly the yellow of the patch on the bottom of the fruit. Cut squash from plants, taking at least an inch of stem.

Cucurbitaceae: Butternut Squash, Cucumber, Green Squash, Hybrid Cantaloupe, Watermelon, Winter Squash; Lettuce

Giant Zucchini Noodles

4 cups zucchini noodles
1 small red pepper
1 tablespoon olive oil
1 tablespoon butter
½ teaspoon lemon juice
¼ teaspoon each of dried basil, marjoram, oregano
2 teaspoons minced garlic

1. Wash and peel the zucchini. Use the vegetable peeler to make roughly 4 loose cups of long ½- to 1-inch "noodles". Chop half a red pepper into roughly ½-inch squares.

2. Add 1 tablespoon of olive oil, 1 tablespoon of butter, and ½ teaspoon of lemon juice to a skillet. After the butter melts, add the noodles and a dash of dried basil, marjoram, and oregano. Cook the zucchini at medium heat for about 4 minutes, turning often.

3. Add the red pepper and 2 teaspoons of crushed garlic. Cook 2–3 minutes until the garlic just begins to turn golden brown. Salt to taste and serve.

It happens to every gardener. Try as you might to harvest zucchinis when they are young and tender, some hide in the vines until they grow to the size of dirigibles. What to do with them?

Short of drying them as firewood, consider making sautéed zucchini noodles. Here's a quick and simple recipe that serves four.

LETTUCE

LETTUCE

TEMPERATURES

Cool. Day temperatures of 60° to 65°F and night temperatures of 50° to 55°F are ideal, although plants can tolerate temperatures close to freezing and as high as 80°F without losing quality.

SOIL

Lettuce requires well-drained, high fertility soil with good nitrogen and humus content, pH of 5.8 to 6.8.

PLANTING

For very earliest crops, start seeds indoors about six weeks before the frost-free date. Place seeds on top of soil mix because light aids germination. Germinate at 55° to 65°F soil temperature. Grow plants at air temperatures of 45° to 50°F nights and 55° to 60°F days. Transplant when plants are 6 weeks old or have four developed leaves, covering them with row-cover material until the weather has settled.

SPACING

For heads, allow 10 to 12 inches between plants in rows 1 foot apart. For leaf lettuces, grow as close together as 1½ inches in rows 4 to 6 inches apart.

CULTURE

Keep young lettuce plants well weeded. Steady moisture is necessary for best growth. If fungal problems develop, thin plants to improve air circulation.

AVERAGE YIELDS

One head per row foot.

HARVEST

For whole heads, cut when plants are full size. If the cultivar is a true head lettuce, gently squeeze it to feel whether it is somewhat solid. Heads of leaf lettuces should simply be dense and close to the size listed in the cultivar description. For "cut and come again" harvests, cut off leaves about 1½ to 2 inches above the crown. Immediately after cutting heads or leaves, plunge them into very cold water, and let them cool for at least half an hour. Shake or spin dry before storing.

SWISS CHARD

MESCLUNS: ARUGULA, ENDIVE, SWISS CHARD

TEMPERATURES
Cool.

SOIL
Fertile, moist, well-drained.

PLANTING
Start seeds in flats, even after weather is warm. Plant each cultivar in a separate container or row in a channel flat. Transplant to the garden, again in separate rows for different cultivars, when seedlings are 3 to 4 weeks old. Plant every 10 to 14 days throughout the season, changing cultivars to match environmental conditions.

SPACING
One-half to ¾ inch apart in the starting container. Break into 2- to 3-inch-long blocks at transplanting time, and set these 4 inches apart in rows at least 8 inches apart.

CULTURE
Keep consistently moist. If you are growing successive crops on the same soil, add compost or a blended organic fertilizer between plantings.

AVERAGE YIELD
One-half pound per row foot.

HARVESTING
Cut leafy crops with scissors when they are 6 inches high, leaving about 2 inches of growth above the crown. Most crops can be cut two times. In the fall or early spring, many hold their quality for a third cutting. Immediately plunge leaves into cold water; drain when they are cool; and spin dry.

OKRA

OKRA

TEMPERATURE
Warm. Okra likes a minimum of 50°F nights, 75° to 85°F days, although plants will tolerate warmer conditions.

SOIL
Soil temperature should be at least 70°F with moderate to high fertility, high humus content, moderately moist, very good drainage, and a pH of 6.3 to 7.0.

PLANTING
In Zones 7 and northward, start seeds early indoors, about four weeks before the frost-free date. If you have fluorescent lights, give seedlings a 14- to 16-hour growing day. Start in individual peat pots or soil blocks. Germinate at 80°F and grow at 65° to 70°F. Transplant to the garden about a week after your frost-free date. In Zones 8 to 11 sow okra directly in the garden.

SPACING
In Zones 6 and north, allow 1 foot between plants in all directions; in Zones 7 and south, allow 1½ feet.

CULTURE
Plants require at least 1 inch of water a week and profit from mid-season side-dressing or foliar feeding, even in good soils. If fertility is low, foliar feed with weak compost tea every two weeks.

AVERAGE YIELDS
One to 2 pounds per plant in average conditions.

HARVESTING
Cut off pods when they are only 3 to 4 inches long for best flavor.

PEAS

PEAS

TEMPERATURES

Cool. Peas like 45° to 50°F nights, 55° to 75°F days.

SOIL

Peas require very deep planting soil with high levels of phosphorus and potassium. If soil is high in nitrogen, plants will produce luxuriant foliage and fewer pods than normal. Conversion of atmospheric nitrogen will also drop. Keep the soil evenly moist, and provide good drainage to prevent root rots. Maintain pH levels between 6.0 to 7.0.

PLANTING

Plant seeds outdoors where plants are to grow as soon as the soil can be worked in spring, generally about four to six weeks before the last expected frost. Plant fall crops about eight to ten weeks before first expected frost date.

SPACING

Plant 2 to 3 inches apart in rows 1 foot apart.

CULTURE

Tall cultivars must be supported on trellises or nets, but even dwarf types yield better if vines are kept off the ground. Push 3-foot-tall twiggy brush into the soil about 6 inches away from the pea row to provide support for these smaller plants. Depending on the characteristics of taller types, they may need supports as high as 5 feet tall.

AVERAGE YIELDS

Two pounds per 10 row feet of English, or shell, peas; 2 pounds per row foot for snap or snow peas.

HARVESTING

Pick snow peas when pods are a few inches long, before they have begun to toughen. Pick snap peas when pods are plump and either before or after peas have begun to fill out, depending on which you prefer. Pick English peas when peas have expanded to fill the pod but before they are so large that they create bulges. When in doubt, hold the pod up to the sunlight to see the peas inside.

RHUBARB

RHUBARB

TEMPERATURES

Hardy from Zones 3 to 8.

SOIL

Rhubarb thrives in deep, rich, moist, high-humus content, well-drained soil with a pH of 5.0 to 6.5. Add at least 2 gallons of fully finished compost to each planting hole.

PLANTING

Plant roots about six weeks before last expected frost in deeply prepared holes. Cover the crown and buds with 2 inches of soil. Once plants are established, roots can be divided as soon as the spring soil can be worked.

SPACING

Plant a minimum of 3 feet apart in rows 6 feet apart.

CULTURE

Keep plants well supplied with moisture in dry periods. In spring and early summer, flower stalks will form. Snap these off before the buds open to allow the plant to conserve energy. In cold climates, mulch the plants for winter with 8 to 12 inches of straw after the top of the soil has frozen.

AVERAGE YIELDS

Yield varies with the age of the plant; fully established plants give 4 to 5 pounds of stems each year.

HARVESTING

As soon as stems are large enough, cut them an inch or so above the crown. Always leave at least 4 to 6 stems on each plant. Cut off all the green leaf material, and compost it. It contains a chemical that is poisonous if eaten but which breaks down in the compost pile.

EGGPLANT

PEPPER

SOLANACEAE: EGGPLANT, PEPPER, POTATO, TOMATO

TEMPERATURES

Eggplant, Pepper, and **Tomato.** Warm. Night temperatures of 55° to 65°F and days of 75° to 85°F are ideal, although eggplants and peppers tolerate warmer days.

Potato. Moderate. Night temperatures of 50° to 60°F and day temperatures of 70° to 80°F.

SOIL

Eggplant, Pepper, and **Tomato.** They prefer well-drained, moderately fertile, humus-rich soil with a pH of 5.5 to 6.8.

Potato. pH of 5.2 to 5.8 is ideal but will tolerate a pH of 6.0 to 6.5.

PLANTING

Eggplant, Pepper, and **Tomato.** Start indoors in almost all locations. Start peppers and eggplants eight to ten weeks before the last expected frost and tomatoes six to eight weeks before the last expected frost. Germinate seeds at a soil temperature of 80°, and grow the seedlings at night temperatures of 50°F or above and days of 65° to 75°F. If using artificial lights, keep them on 12 hours per day.

Potato. The potato itself is the "seed" of the plant. The tuber is actually an enlarged stem, and the eyes are buds from which new stems, roots, and potatoes will grow. Cut seed potatoes into pieces about the size of an egg, each with at least one, but preferably two, eyes. Let the cut portion callus over for a day or two to prevent rotting. Many people plant as much as a month before the last spring frost, but this early planting can lead to fungal diseases, and growth is slow until spring temperatures settle. If fungal diseases have been a problem, wait to plant until the soil is at least 50°F.

SPACING

Eggplant, Pepper, and **Tomato.** Space 2 to 2½ feet apart in rows 2 to 2½ feet apart, depending on size of cultivar.

Potato. Plant pieces about 1 foot apart in rows 2 to 3 feet apart. Plant only 3 to 4 inches deep to avoid rotting.

CULTURE

Eggplant, Pepper, and **Tomato.** Transplant seedlings to the garden when night temperatures are consistently 50° or above, although tomatoes can tolerate night temperatures of 45°F. If this is impossible, use plastic or black paper mulch with poly tunnels or row covers to increase soil temperatures. Supply steady moisture levels while fruits are forming to keep the flavor mild and sweet. Do not over-fertilize, particularly with nitrogen-bearing materials, or plants may produce more leaves than fruit. Monitor for pests and diseases every few days, checking under leaves and on growing tips. Once tomato plants are established and summer has come, keep soil temperatures cool by covering soil or black mulches with a layer of straw. Remove suckers on indeterminate plants (see opposite) every three to four days, also tying them to supports as necessary.

Potato. Tubers form only above the seed potato, so provide lots of space for them to develop. As the plants grow, heap soil from the pathways over them, covering all but a few inches of the top of the plant. Continue to hill the plants this way every week to ten days until midseason. Straw mulches can be used in place of the soil as long as they really cover the tubers. Tubers that are exposed to light will develop the green-colored solanine, a chemical that can make you sick. Steady

Solanaceae: Eggplant, Pepper, Potato, Tomato

POTATO

TOMATO

moisture is important to potatoes; irrigate in dry periods. To maintain the high fertility levels potatoes require, drench the soil around each plant with compost tea or a seaweed/fish emulsion in late June and again in mid-July. Foliar sprays every two to three weeks with a weak compost tea help to prevent some of the worst fungal diseases and give needed nutrition.

AVERAGE YIELDS

Eggplant, ¾ to 1 lb. per row foot. **Pepper,** five to seven fruit per plant for bell peppers, variable for hot peppers. **Potato,** 4 to 6 lbs. per plant. **Tomato,** 1½ lbs. per row foot at a minimum.

HARVEST

Eggplants. Cut when they look full size and still have shiny skin. (Once the gloss begins to dull, the meat is bitter-tasting.) Cut with at least an inch of stem to guarantee good storage.

Pepper. Pick the first pepper on each plant just a bit early to promote more flowering and fruit set. Cut peppers, taking some stem, to prevent harming the somewhat brittle plants. To get high yields of red or other colored bell peppers,

allow the fruit on half your plants to begin to mature once a few green peppers have been harvested from them. Avoid some rots by picking these fruits when they are 50 to 75 percent colored. They will finish coloring if you hold them in dark conditions at 45° to 50°F.

Potato. When potato plants bloom, tubers are forming. If you are careful not to greatly disturb the plant, you can dig directly under it two weeks after it has flowered and pull out a couple of small new potatoes. Wait to harvest the balance until the foliage dies back in the late summer or fall. To avoid some

fungal diseases that develop in storage, dig the potatoes when the soil is moderately dry, before the first frost. Do not wash the potatoes, but set them in a dark spot to let the soil on the skin dry before storing.

Tomato. Pick fruit when it is at least half-colored for a vine-ripened flavor. Many cultivars split if left to ripen fully on the vine, particularly if it rains during their last day or two of coloring. Set fruit in a dark, warm place to finish ripening. Avoid leaving overripe or rotting fruit on the vine or ground because disease spores will travel through your planting.

Tomatoes: Determinate or Indeterminate?

Tomatoes can be "determinate," meaning that they are bred to grow to a certain height, or "indeterminate," meaning that they will continue to grow for as long as they are alive. Catalogs indicate which is which.

Knowing how a tomato plant grows is important. Determinate tomatoes form fruit on the branches (or suckers) that grow

between the main stem and first branches. If you remove these suckers, you will be removing the fruit-production sites. In contrast, removing suckers is a good way to prune indeterminate plants. Because the plants continue to grow from the top of the stem, they continue to produce blossom- and fruit-bearing branches.

Garden-Bounty Salad

When your garden is really cranking you may find you have too much of a good thing—especially tomatoes and cucumbers. Here's a simple but delicious salad you can put together in minutes. Make it the night before serving for full flavor. Bountifully serves eight.

1. Chop forkful size chunks of the following: tomatoes, 4 cups; cucumbers, 4 cups; green peppers, 1½ cups; sweet onion, 1 cup.
2. Place the above in a large bowl, and add: garlic, 2 or more cloves, crushed; ½ cup red wine vinegar; ¼ cup olive oil.
3. Mix thoroughly, adding salt and fresh-ground pepper to taste. Chill.

SPINACH

SPINACH

TEMPERATURES

Cool. Night temperatures of 45° to 50°F and day temperatures of 60° to 65°F are ideal.

SOIL

Spinach likes soil high in nitrogen, well-drained but moist, with a pH of 6.5 to 7.5.

PLANTING

Start seeds outdoors in very early spring where plants are to grow as soon as the soil can be worked. Use row covers to protect plants from freezing temperatures. Plant every seven to ten days, scheduling so that plants will mature before temperatures are consistently above 75°F. Fall crops can be planted six weeks before the first expected fall frost.

SPACING

Plant seeds 1 to 2 inches apart in rows 1 to 1½ feet apart. Thin plants as they grow, removing alternate plants until remaining plants stand 8 to 10 inches apart.

CULTURE

Weed young plants early and regularly, and keep the soil consistently moist. If leaves look yellow, foliar-feed with compost tea or a solution of liquid seaweed/fish emulsion. Keep plants thinned so that leaves of adjacent plants do not touch, because crowding also promotes early bolting.

AVERAGE YIELDS

Two pounds per row foot.

HARVESTING

Thin alternate plants to 8 to 10 inches apart. To harvest whole plants, cut the main stem just below the crown so that leaves stay intact on the plant. To prolong the harvest of fall crops, break off bottom leaves as you need them. Immediately cool plants or leaves in very cold water for ½ hour, and shake dry before refrigerating.

SWEET POTATO

SWEET POTATO

TEMPERATURES

Warm. Night temperatures no lower than 55°F and day temperatures between 80° to 90°F are ideal.

SOIL

Average fertilty, high humus levels, moist but well-drained, with a pH of 5.5 to 6.0. Calcium, magnesium, and boron should all be in balance. If not, apply foliar spray of seaweed/fish emulsion and liquid seaweed every four weeks.

PLANTING

Sweet potatoes are started from "slips." Purchase them in the spring from seed companies. You can grow your own slips indoors six to eight weeks before the frost-free date by simply laying the tubers on good garden soil and covering them with straw. Choose a reliably warm cold frame, greenhouse, or even a windowsill. When the slips are about 6 to 8 inches long, break them off and root them in a flat of fresh potting soil. In about a week, when they have roots, plant them outdoors.

SPACING

In Zones 6 and colder, plant slips 1 to 1½ feet apart in rows 2 to 3 feet apart. In Zones 7 to 11, allow 3 feet in all directions. Sweet potatoes can be trained up trellises in all climates.

CULTURE

In Zones 4, 5, and colder, black plastic mulch and row-covers or slitted poly tunnels raise temperatures and increase yields dramatically. In all climates, keep plants weed-free when they are young. Once they become established, the plants will shade out all possible competition.

AVERAGE YIELDS

Fifteen to 30 tubers per plant in optimum conditions.

HARVEST

Dig by hand. Choose a warm, dry day a week or so before you expect the first frost.

CARROT

UMBELLIFERAE: CARROT, CELERY, PARSNIP

TEMPERATURES

Moderate. Night temperatures of 50 to 55°F and day temperatures of 65° to 80°F. Mature roots withstand freezing. Temperatures above 80°F make plants stringy.

SOIL

Deep, sandy or loose, well-drained, fertile, moist, and rock-free soil with a pH of 5.5 to 6.8 is ideal.

PLANTING

Carrots. Wait until the threat of severe frost has passed, and plant successively every three or four weeks until three months before expected fall frost date.

Celery. Start seed early indoors about 10 to 12 weeks before the frost-free date. Seeds may take several weeks to germinate, even at ideal soil temperatures. Plants are slow-starting but must not be allowed to dry out or get too cool during their seedling phase. Transplant to the garden once all danger of frost has passed. If nights threaten to dip below 55°F, protect plants.

Parsnips. Use only fresh parsnip seed because it loses viability quickly. Soak seed overnight in weak compost tea or liquid seaweed, diluted to half the strength recommended on the bottle. Direct seed in spring or early summer once the soil is 55°F or above. Cover seed furrows with vermiculite and row-cover material. Seed will take three to four weeks to germinate. In Zones 4 and northward, seeds can be started indoors in 3-inch peat pots about six weeks before the frost-free date.

SPACING

Plant **carrot** and **parsnip** seeds ½ inch apart in rows 12 inches apart. Thin to 2 inches apart as you weed. For **celery,** leave 8 to 12 inches between plants in rows 1½ to 2 feet apart.

CULTURE

All umbelliferae must be kept as weed-free as possible, but this is particularly true for the root crops. If possible, wilt the weeds before seeds sprout using a propane flame weeder. Keep rows well watered all through the season, and beginning in mid-season, side-dress with compost; drench with a cup of compost tea for every row foot of growing area; or foliar-feed with a seaweed/fish emulsion every month.

AVERAGE YIELDS

Carrot, 1 to 1½ pounds per row foot. **Celery,** 2 to 3½ pounds per row foot.

HARVEST

Carrots. They develop their sugar content during the last three weeks before they are mature. Pull and taste-test a carrot or two to determine whether they are ready to be harvested.

Celery. Cut just below the soil surface. Cool in cold water immediately. To prolong the harvest of individual plants, snap off one or two outer stems as needed.

Parsnip. Parsnips taste best after hard frosts. Wait to dig until there have been at least two frosts, one of which has been heavy. Use a spading fork to loosen the soil along the sides of the row, and then grasp the top of the root and pull.

Raising Vegetables and Herbs

Gardening through the Seasons: Spring

Spring is a busy time, beginning with spreading compost on the beds, working the soil, and planting. The planning and seed selection you've done through winter comes to fruition as you get your crops started. It is also a time for erecting trellises, bean tepees, and any support stakes before plants get too large—do so later, and you might disrupt roots. If you have a drip irrigation system, spring is time to rearrange it to suit your crop rotation.

Spring can bring surprises such as a late freeze or snowfall that can jeopardize your young plants and sets. Have plastic handy to protect the plants.

Early seedings include radishes, lettuce, and sets for Brussels sprouts and cauliflower.

It's time to put that old patio to use and try a straw-bale garden. First step is soaking the bales for several days.

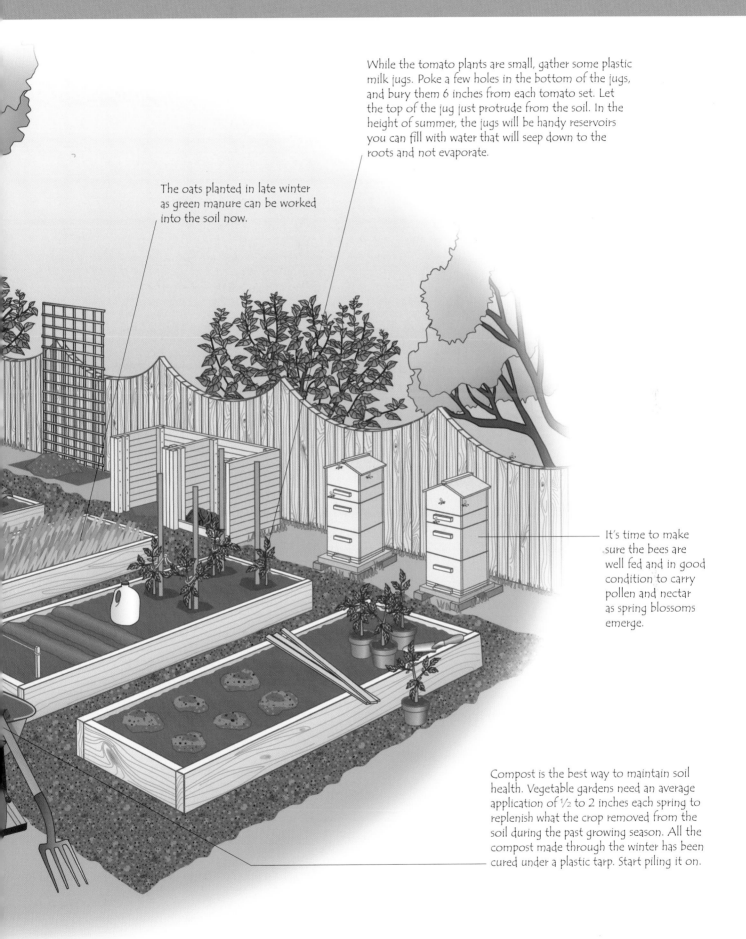

While the tomato plants are small, gather some plastic milk jugs. Poke a few holes in the bottom of the jugs, and bury them 6 inches from each tomato set. Let the top of the jug just protrude from the soil. In the height of summer, the jugs will be handy reservoirs you can fill with water that will seep down to the roots and not evaporate.

The oats planted in late winter as green manure can be worked into the soil now.

It's time to make sure the bees are well fed and in good condition to carry pollen and nectar as spring blossoms emerge.

Compost is the best way to maintain soil health. Vegetable gardens need an average application of ½ to 2 inches each spring to replenish what the crop removed from the soil during the past growing season. All the compost made through the winter has been cured under a plastic tarp. Start piling it on.

Raising Vegetables and Herbs

Gardening through the Seasons: Summer

Summer is dedicated to making sure all of your crops are doing well. That means making sure they have enough water, controlling weeds, and watching for disease and insect damage. It is also the time to check that flourishing plants are supported, keeping tomatoes tied to their stakes or watching that cages don't get overwhelmed. It is a time of harvest too: radishes, snow peas, rhubarb, herbs, and strawberries can be gathered in as they produce.

Beans, squash, cucumbers—even pumpkins—are just a few of the crops that benefit from being trained so that they don't grow along the ground where they'll be prone to pick up diseases and insect pests. Trellising improves air circulation, keeps animal and soil-dwelling pests away from the crops, and decreases soil-borne diseases. Training tomatoes up a trellis or supporting them with a cage or pole instead of letting them sprawl has a huge impact on their health.

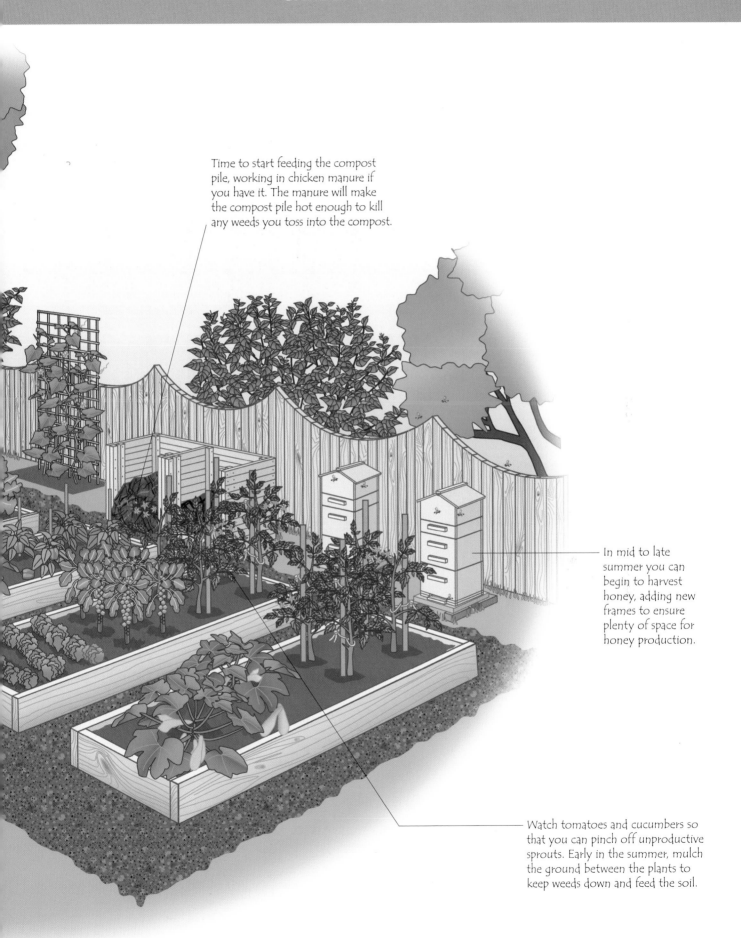

Time to start feeding the compost pile, working in chicken manure if you have it. The manure will make the compost pile hot enough to kill any weeds you toss into the compost.

In mid to late summer you can begin to harvest honey, adding new frames to ensure plenty of space for honey production.

Watch tomatoes and cucumbers so that you can pinch off unproductive sprouts. Early in the summer, mulch the ground between the plants to keep weeds down and feed the soil.

2 Raising Vegetables and Herbs

Gardening through the Seasons: Fall

Fall is harvest time, another busy period as you bring in the crops, help some plants extend their productive life into the early winter, and build up compost. It is also time to clean things up, ridding the beds of plant matter that might harbor pests and diseases.

If you have herbs you have been enjoying through the summer, make a final harvest for drying.

Dig up your potatoes before the first fall frost.

Gather up poles and cages after harvest, and clean them up so they're ready in the spring. Keep the pumpkins healthy by putting a square of tarpaper under them.

Mulch all your perennials with straw to protect them through the winter.

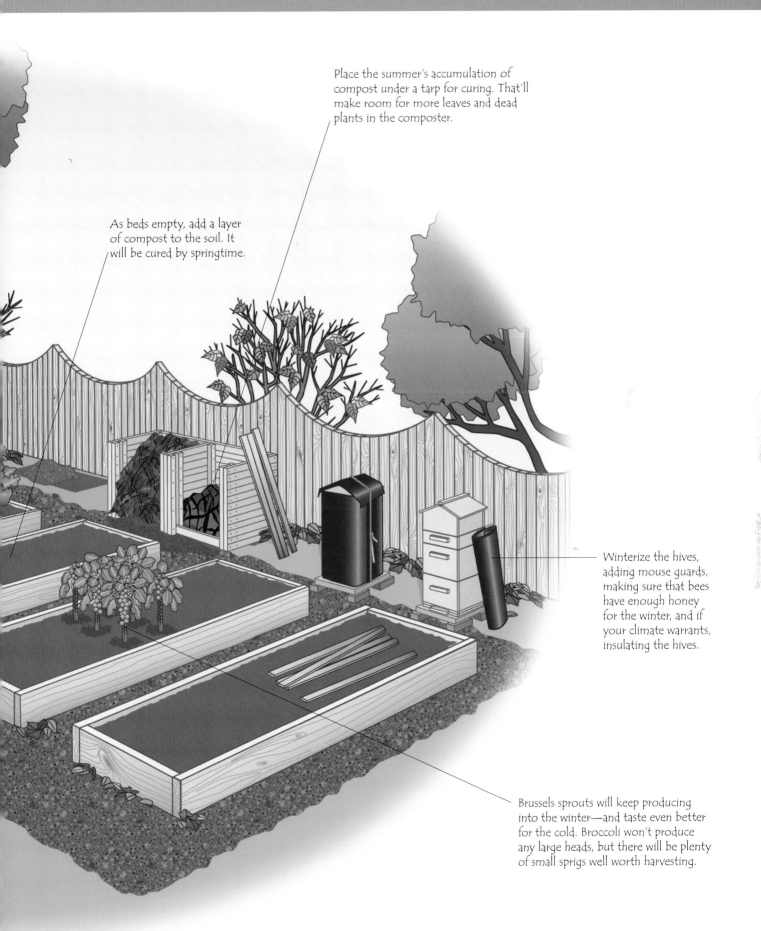

Place the summer's accumulation of compost under a tarp for curing. That'll make room for more leaves and dead plants in the composter.

As beds empty, add a layer of compost to the soil. It will be cured by springtime.

Winterize the hives, adding mouse guards, making sure that bees have enough honey for the winter, and if your climate warrants, insulating the hives.

Brussels sprouts will keep producing into the winter—and taste even better for the cold. Broccoli won't produce any large heads, but there will be plenty of small sprigs well worth harvesting.

Herbs

Herbs are so easy to grow and have so few pest and disease problems that they rarely create extra work. If you've priced fresh herbs lately, you'll know that growing your own can save you a bundle. Many people grow herbs among vegetables or mixed into flower beds rather than in separate herb gardens. This method allows you to take advantage of herbs' benefits in places where they are most needed.

Herbs make ideal container plants, too. If your outdoor space is limited, you may want to plant your herbs in pots. Containers can complement the looks of the herbs and give you maximum flexibility for location.

Most herbs have a long period of usefulness. Unlike most vegetables and fruits, many perennial herbs can be picked at any time during the growing season. Some are generally picked before or after they bloom, but they can still be used while they are flowering. Annual herbs taste strongest before they bloom. Experienced gardeners work around this by picking consistently to prolong the life of the herbs as well as making second and sometimes third plantings of the fastest-maturing species.

When you plan an herb garden, you may think first of the many ways you can use herbs in cooking. But once you start growing them, you'll have a large enough supply to begin experimenting with other uses. Herbal preparations are amazingly easy to make. From herbal vinegars to potpourris, infused oils, salves, and skin creams, all it takes to create these treasures is a little time and a few simple ingredients and kitchen tools.

Close to the Kitchen

It's a treat to have dill, basil, tarragon, and chervil just outside the door. So whether it's a container garden, or a bed in the ground, be sure to plant some kitchen herbs to use in your cooking.

In addition to their beauty and rich flavors, herbs are also low-maintenance crops.

Grow vertical. Use the areas that face the sun along fences and walls to place containers of herbs and vegetables. Stacked planks, makeshift shelves, and even an old trellis like the one shown right, which is used as a base for hanging containers, can support a variety of crops.

Kitchen herb gardens are ideal places to mix perennial and annual plants. Site the perennial herbs first, leaving ample space between them so that you can tuck in annuals during the season. Tarragon, oregano, some of the thymes, mints, chives, and sage usually form the perennial backbone of the kitchen garden. Oregano and the thymes make good edging plants, while sage, chives, and tarragon grow large enough to be used in the middle or back of the bed. Confine mints to pots or a garden all their own. Successive plantings are the best way to maintain steady supplies of many annual herbs.

Corralling Mint

Mint spreads, so to confine it, divide the area into squares that measure at least 2 feet on each side. Plant one mint in the center of each of these areas, and mulch the space with straw or another nonweedy material. At monthly intervals during the summer, pull back the mulch, and use a sharp spade to cut into the soil at the divisions of the squares. By the second year, you'll have to prune off traveling stems, which, along with underground rhizomes, will be trying to move into their neighbors' areas.

Containers

Herbs are wonderful container plants. The qualities that make container gardening a bit more demanding than growing in the soil are often inconsequential or even beneficial for many herbs. For example, the soil in containers dries out quite quickly. Fast-growing annual herbs, such as basil and chervil, require as much water as vegetable plants. But the heat-loving Mediterranean herbs—oregano, thyme, and rosemary—all welcome a bit of drying between watering.

Vegetables in containers normally require fertilization every week to ten days. Again, this will be true for some of the fast-growing annual herbs, but perennials such as tarragon, rosemary, thyme, oregano, and sage are less demanding. They require supplemental feeding, but less frequently. Use a mixture of liquid seaweed and fish emulsion once a month. If the leaves begin to yellow, fertilize container-grown herbs once every two weeks.

Small containers can stunt plants. When plants become root-bound, they slow their top growth—not something you'd want for vegetables. However, given that a rosemary can grow into a shrub within two or three years, a single sprig of lemongrass can become

A modest amount of a herb goes a long way. That, plus some herbs' propensity to take over a plot, make herbs good candidates for containers.

a 3-foot-wide thicket in a single summer, and oregano and thyme can take over an area several feet in all directions, selective dwarfing is a good thing.

Common Herbs, Plant by Plant

SWEET BASIL

SWEET BASIL

Life Cycle: Annual.

Size: 1 to 2 feet tall, 8 to 12 inches wide.

Exposure: Full sun.

Soil: Well-drained, moist, nutrient-rich.

Propagation: Start seeds early indoors, about five weeks before the frost-free date, or seed directly in

Use fresh or dry leaves as a seasoning in composed dishes, the primary ingredient in pesto, or to infuse vinegars or oils. Florets can be added to salads or used as a garnish. Bees love basil. Transplant or direct-seed a second crop in early to mid-July to ensure a continuous supply of leaves throughout the season.

the garden after all danger of frost has passed.

Spacing: Allow 1 to 1½ feet between plants.

Harvesting Guidelines: Begin pinching off branch tips when plants are 6 to 8 inches tall. Always leave a pair of leaves on the stem because new leaves will grow from the buds in the leaf axils. If flowers do form, pinch them off to prolong the usefulness of the plant. Harvest only when plants are dry. If storing for a day or so, stand the stems in an inch or so of water, and keep cool and dark.

Preserving Techniques: Dry by hanging or in electric dehydrator. Blanch and freeze in ice cubes or on cookie sheets. Freeze raw in butter or oil.

CHERVIL

CHERVIL

Life Cycle: Annual.

Size: 1 to 1½ feet tall, 6 to 8 inches wide.

Exposure: Prefers partial shade but will tolerate full sun in the North.

Soil: Moist, rich soil with high levels of organic matter.

Propagation: Plant seeds where they are to grow a week or two before the last expected frost. If

Add this lightly licorice-flavored leaf to soups, salads, casseroles, and peas immediately before serving. In France, it is often used in preference to parsley. Many beneficial insects feed on the flowers.

plants set seed, they give a volunteer crop late in the season or the following year. Plant chervil every two to three weeks during the season to have a steady supply. For midseason plantings, choose a location that is shaded during the heat of the day.

Spacing: Allow 6 to 8 inches between plants.

Harvesting Guidelines: Pinch leaves when plants are 6 to 8 inches tall. Crop can be extended if stems are picked before the plant forms blooms. Harvest only when plants are dry.

Preserving Techniques: Chervil can be frozen in ice cubes or butter but loses flavor if dried.

Sweet Basil; Chervil; Chives, Chinese or Garlic Chives; Coriander

CHIVES

Use chives in salads, soups, or any other dish where a mild onion or garlic flavor is needed. Flowers of both kinds of chives are good additions to salads and make pretty garnishes. Bees love chive flowers.

CHIVES, CHINESE OR GARLIC CHIVES

Life Cycle: Perennial.

Size: Both plants grow 1 to 1½ feet tall and eventually form clumps measuring 8 inches to 1 foot wide.

Exposure: Prefers full sun but will tolerate partial shade.

Soil: Well-drained, moderate fertility with high humus content.

Propagation: Start seeds indoors eight to ten weeks before frost-free date, and transplant to garden a week before that date. Seeds can also be directly sown, but plants can be mistaken for grassy weeds. Garlic chives are particularly offensive to Japanese beetles, so many people grow them with roses and other plants that are plagued by these insects.

Spacing: Plant about ten seeds in each pot or cell of the starting flat. Space clumps of plants about a foot apart in the row. Plants will enlarge to fill in the row.

Harvesting Guidelines: When plants are 5 inches tall, cut leaves 2 or 3 inches above the soil surface. Cut flowers just after they open. Harvest only when plants are dry.

Preserving Techniques: Snip chive stems into small pieces, and freeze in ice cubes. Drying is rarely successful.

CORIANDER

Salsa, chili, and many curry dishes rely on coriander and its leaves, called cilantro. Crushed seeds give excellent flavor to safflower or canola oil. Many beneficial insects feed on the flowers.

CORIANDER (CILANTRO)

Life Cycle: Annual.

Size: 2 to 3 feet tall, 6 to 8 inches wide.

Exposure: Prefers full sun in the North and partial shade in the South.

Soil: Well-drained, moderate fertility with good moisture levels.

Propagation: Plant seeds in early spring where plants are to grow, or use peat pots or soil blocks to avoid disturbing roots when you transplant. Plant a new crop every 2 to 3 weeks through the season to have a steady supply. If you allow seeds to drop to the ground, volunteer plants will appear late in the season or the next year.

Spacing: Place 8 to 10 inches apart in the row.

Harvesting Guidelines: Begin cutting leaves (cilantro) when plants are 6 to 8 inches tall and continue for as long as possible. To prolong the harvest of leaves, cut the flower stalk as soon as it forms. To harvest seeds, allow flower stalks to form and insects to pollinate. As the flowers begin to dry, watch them carefully. When they appear almost dry, place a paper sack over them and secure it at the bottom with a twist-tie. Bend the seed head over and shake. When you can hear the seeds drop into the bag, cut the stalk and finish drying the seed head inside.

Preserving Techniques: Coriander does not dry well. Freeze leaves in ice cubes or use to flavor oils.

DILL

DILL

Life Cycle: Annual.

Size: 1½ to 5 feet tall, depending on cultivar; 8 to 12 inches wide.

Exposure: Full sun.

Soil: Well-drained, moist, and fertile, but will tolerate leaner soils.

Propagation: If starting seeds early indoors, plant groups of four or five seeds in peat pots or soil blocks and space these clumps a foot apart when you transplant. Otherwise, in early spring, plant seeds where the plants are to grow.

Culinary-leaves and seeds are used to flavor soups, cheeses, and a variety of mixed dishes as well as dill pickles. Seeds and leaves can flavor vinegars or oils. Many beneficial insects feed on the flowers.

Spacing: Allow 6 inches between plants unless they are grown in clumps, as described above.

Harvesting Guidelines: Clip off individual leaves beginning when plants are 8 inches tall. If cutting to dry, allow plants to reach close to full height and then cut before flowers form. To harvest seeds, allow plants to flower and set seed. Enclose seed heads in plastic bags once they're nearly dry.

BAY LAUREL

BAY LAUREL

Life Cycle: Perennial in Zones 8 to 10.

Size: 5 to as much as 60 feet tall in its native environment; 3 to 15 feet wide.

Exposure: Full sun.

Soil: Well-drained, moderately rich, consistently moist but not soggy.

Propagation: Buy established plants, and propagate by taking stem cuttings in the fall. Seeds usually rot. In northern areas, move your potted bay tree in and out, taking care to bring it back inside a few weeks before the first fall frost.

Spacing: Plan for at least 15 feet between trees.

Harvesting Guidelines: Pick off leaves or whole stems when the plant is a foot tall.

Preserving Techniques: Hang upside down to dry. Weigh leaves with heavy objects to keep them from curling.

Use bay leaves in a variety of cooked dishes. Plant a bay tree in a container if you live in Zones 7 and north, and enjoy it indoors.

Dill; Bay Laurel; Lemon Balm; Lemongrass

LEMON BALM

LEMON BALM

Life Cycle: Perennial, Zones 4 to 9.

Size: 1 to 2 feet tall.

Exposure: Full sun in the North; filtered shade in the South.

Soil: Moist, well-drained, with average fertility.

Propagation: Start seeds early indoors, approximately eight weeks before the last expected frost. Transplant outside after all danger of frost has passed. Or you can plant seeds where they are to grow, about two weeks before the frost-free date. Stem cuttings can be taken from established plants at any time during the growing season.

Spacing: Allow approximately 2 feet between plants.

Harvesting Guidelines: Cut off stem tips as needed. When cutting lemon balm for drying, cut stems about 4 inches above the soil surface.

Preserving Techniques: Strip leaves from the stems, and dry them on the fruit leather insert of a food dehydrator. Freeze whole leaves in ice cubes to use in iced teas and other cold beverages.

Lemon balm leaves can be minced and used fresh or cooked in a variety of dishes. It is also is one of the best tea herbs. Lemon balm won't invade the garden as the other mints do.

LEMONGRASS

Leaves and bulbous stems are commonly part of the Asian diet and are used in a huge number of Thai and Vietnamese dishes. Dried leaves make an excellent addition to an herbal tea mixture. Ornamental-dried leaves add a pleasant lemony fragrance to potpourri.

LEMONGRASS

Life Cycle: Perennial, Zones 9 to 10; grown in containers elsewhere.

Size: 3 to 5 feet tall; 1 to 3 feet wide.

Exposure: Full sun.

Soil: Moist, rich soil with a pH of 6.5. In containers, fertilize once a month during the season.

Propagation: Start seeds in early spring, and transplant them to your garden or a larger container after all danger of frost has passed. Clumps enlarge naturally and quickly, so divide plants every year or so. When dividing the plants, cut back the leaves to between 3 and 4 inches above the soil surface. If you live in the North, you can grow lemongrass in the ground during the summer. Remember to dig it out and pot it so that you can bring it back inside several weeks before your first expected frost and before you turn on the heat. This will give the plant ample time to adjust to the warmer, drier conditions in the house.

Spacing: Allow 2 to 4 feet between plants in Zones 9 to 10. Elsewhere, grow plants in 5-gallon nursery pots through the winter.

Harvesting Guidelines: Once the diameter of the base is about 2 inches wide, pull off older outside bulbous stems from the base. In this way you'll be taking the leaves, too. You can also cut the leaves from the plant for use in a tea without having to take the bulb.

Preserving Techniques: Use fresh because the plant does not retain full flavor when dried.

LEMON VERBENA

LEMON VERBENA

Life Cycle: Perennial, Zones 9 to 10, container plant elsewhere.

Size: 2 feet tall by 2 feet wide.

Exposure: Full sun.

Soil: Well-drained, nutrient-rich soil with high humus content.

Propagation: Buy established plants. In late spring or early summer, take cuttings and keep them in a high humidity area while they are rooting.

Spacing: Outdoors, allow 2 feet between plants. Plant in 12-inch pots.

Harvesting Guidelines: Snip leaves as needed. When the plant begins to look scraggly, cut all the branches back to 4 inches and let them regrow.

Preserving Techniques: Dry on screens or in an electric dehydrator.

Use the leaves in salads, desserts, sauces, and dressings. Lemon verbena makes an excellent tea, alone or in combination with other herbs. Pinch a leaf of lemon verbena for an instant air freshener. Indoors grow the plant under a wide-spectrum plant light to keep it healthy. Grow lemon verbena in hanging baskets or containers near the outdoor living area.

MINT

Use mint in salads and with stews, fish, grains, and peas. Most mints make excellent tea. Many beneficial insects feed on the tiny mint flowers when they are in bloom.

MINT

Life Cycle: Perennial, Zones 5 to 9.

Size: 1½ to 2 feet tall, 1 to 3 feet wide.

Exposure: Partial shade but will tolerate anything from full sun to shade.

Soil: Moist, moderate fertility. Do not plant in high-nitrogen soils.

Propagation: Most cultivars do not come true from seed and must be propagated from cuttings or divided. Take cuttings before the plants bloom. Divide plants in early spring or early fall. If you do plant seeds, expect to get mint plants that are fine for tea, but which contain less mint flavor than the named cultivars. Mints deserve their reputation for being invasive. Do not plant them in a mixed bed or they will take it over within only a couple of years. Instead, plant one or more species in a bed or area by themselves. Even so, you'll have to watch to see that one or another type doesn't become dominant.

Spacing: Allow 1 foot in each direction.

Harvesting Guidelines: Pinch off stems, tips, or individual leaves from dry plants as needed. If harvesting to dry, cut stems before the plant blooms and then again in the fall. Cut 4 to 6 inches above the soil surface for the first cutting and only an inch above it in the fall.

Preserving Techniques: Hang stems upside down to air dry or strip leaves from stems and dry in an electric dehydrator. Pack dried leaves in glass jars or resealable plastic bags for storage. Freeze leaves in ice cubes or chopped in butter or oil.

Lemon Verbena; Mint; Nasturtium; Oregano

Use flowers and leaves in salads and as edible garnishes. Pickle seedpods in heated vinegar to use as you do capers.

NASTURTIUM

Life Cycle: Annual.

Size: 1 to 2 feet tall, from 1 to 6 feet wide, depending on cultivar.

Exposure: Full sun.

Soil: Moderate fertility, well-drained.

Propagation: Start seeds early indoors in 3- to 4-inch peat pots four to five weeks before the last expected frost. Transplant to the garden or move outdoors after all danger of frost has passed. Seeds can also be planted where they will grow after all danger of frost has passed.

Spacing: 1 foot apart in all directions.

Harvesting Guidelines: Once plants are 8 to 10 inches tall, pinch off leaves and flowers as needed through the season.

Preserving Techniques: Nasturtium leaves and flowers do not preserve well; use fresh.

A Nitrogen Tell-Tale

Nasturtiums will indicate how much nitrogen is in your soil. If the leaves are only 1 to 2 inches across, the soil is nitrogen deficient. If they are 3 to 4 inches across, nitrogen is low for most vegetables but will support light-feeding herbs and flowers. Leaves that are 6 inches across indicate that nitrogen-loving vegetables can grow well.

Use oregano in sauces, pizza, and other cooked dishes.

OREGANO

Life Cycle: Perennial, Zones 5 to 9.

Size: 1 to 2 feet tall, 1 to 2 feet wide. Stems are usually recumbent so the plant appears only a few inches tall.

Exposure: Full sun.

Soil: Well-drained, average fertility.

Propagation: Start seeds early indoors, about eight to ten weeks before the frost-free date. Plant seeds in groups of two or three. Transplant to the garden just before the frost-free date. You can also layer plants covering the stems with soil and allowing them to form roots. Plants will propagate themselves after the first year and the plot will need thinning. Oregano plants vary in flavor. If you have one you like, propagate it by layering. If you are starting from scratch, look for 'Greek' oregano with white flowers. Pink flowers signal that the flavor and holding quality is inferior.

Spacing: Allow a foot between groups, 6 inches between individual plants.

Harvesting Guidelines: Pinch off stem tips as needed once the stems are at least 6 inches long. If harvesting for drying, cut whole stems just before the plant flowers. A second cutting may be made in late August. When you take stems, cut 2 to 3 inches above the soil.

Preserving Techniques: Strip leaves from stems. Dry on the leather insert of an electric dehydrator, or hang stems upside down over a clean piece of cloth.

PARSLEY

Life Cycle: Biennial but usually treated as an annual.

Size: 10 to 15 inches tall, 8 inches to 1 foot wide.

Exposure: Full sun in the North, partial shade in the South.

Parsley is used as a garnish, an ingredient in many cooked dishes, and as a primary component of some pestos.

ROSEMARY

Life Cycle: Perennial, Zones 8 to 10, grow in container elsewhere.

Size: 2 to 6 feet tall, 1 to 4 feet wide.

Lamb and poultry seasoning wouldn't be the same without rosemary. Add rosemary to sachets and potpourris for a clean, piney scent.

SAGE

Life Cycle: Perennial, Zones 4 to 8.

Size: 2½ to 3 feet tall, 1 to 2 feet wide.

Exposure: Full sun.

Soil: Well-drained, average fertility.

Fresh and dried leaves are used in many cooked dishes. Sage stems make a good base for herbal wreaths.

Parsley; Rosemary; Sage

Soil: Well-drained, moist, high fertility.

Propagation: Start seeds early indoors, about eight to ten weeks before the frost-free date. Germination can take as long as three weeks. To speed it up, soak seeds for 24 to 30 hours. Drain them in a tight-meshed strainer. Every few hours for the next two to three days, run tepid water over the seeds in the strainer to wash off the compound that inhibits germination. At night, put the seeds back in a jar or sealed plastic bag, along with a wet cotton ball, to keep them from drying out. Seeds will usually germinate in five days after this treatment. Plants that overwinter in the garden will bloom in late spring of their second year. To bring in for the winter, pot in a container 1 to 1½ feet deep and supplement light from a south window with a grow light.

Spacing: Allow 1 foot in each direction.

Harvesting Guidelines: Cut outside stems first, about 1 inch above the soil surface.

Preserving Techniques: For the best flavor, freeze stems on cookie sheets and then pack them in plastic bags, or mince and freeze in ice cubes. Air-dried parsley loses flavor and color, but it will hold both if you dry it in an electric dehydrator.

Exposure: Full sun.

Soil: Well-drained, moderately fertile soil. Do not overwater.

Propagation: Rosemary can be started from seed, but most people take cuttings from established plants in early spring or late summer. Stems may also be layered in late spring or early summer.

Spacing: Allow 1 to 3 feet between plants, depending on size. North of Zone 8, plant rosemary outside once all danger of frost has passed. Several weeks before the first expected fall frost, dig it up and pot it in a deep nursery container for the winter. Overwinter it in a south window or under plant lights. Plants will live for many years, eventually becoming too large to transplant each year. Instead, grow them in 10-gallon nursery tubs on wheels so you can move them.

Harvesting Guidelines: Once plant is 6 to 8 inches tall, snip off branch tips as needed. If harvesting to dry, cut green stems to within a node or two of the previous year's woody stems.

Preserving Techniques: Dry flat on screens or in an electric dehydrator.

Propagation: Start seeds early indoors, about eight weeks before the frost-free date. Freeze seeds for a week before planting. Cover seedling flats with newspaper; they require darkness to germinate. Transplant to the garden a week before or after the frost-free date. Or take softwood cuttings in spring or early summer. Sage plants lose their good looks sometime in the third or fourth year. If you want them to do double-duty in your garden, providing beauty as well as flavor, make new plantings every third year and remove the old plants. Change the location of the new plants to avoid any root diseases that may have built up in the area.

Spacing: Place 2 feet apart in all directions.

Harvesting Guidelines: Snip branch tips as needed. When harvesting stems, cut a node or two above the woody growth, preferably in late summer or early fall.

Preserving Techniques: Hang stems upside down to air-dry.

SWEET MARJORAM

SWEET MARJORAM

Life Cycle: Perennial grown as annual.

Size: 1 foot tall, 10 inches wide.

Exposure: Full sun but tolerates partial shade.

Sweet marjoram tastes like a sweet oregano. Use it raw or cooked whenever you want a mild oregano flavor.

TARRAGON

TARRAGON, FRENCH TARRAGON

Life Cycle: Perennial, Zones 4 to 8.

Size: 1½ to 2 feet tall, 1 to 1½ feet wide.

Tarragon is used with poultry and fish, and also makes a good addition to salad dressings and many sauces. Make tarragon vinegar with high quality white wine vinegar.

THYME

THYME

Life Cycle: Perennial, Zones 5 to 9.

Size: 6 inches to 1¼ feet tall, 6 inches to 1¼ feet wide, depending on cultivar.

A delicious complement to garlic, thyme is ideal for lamb, pork, or beef roasts. It also seasons cheese, tomato, and egg dishes. Many beneficial insects are attracted to thyme flowers, but the dried leaves can act as an insect repellent.

Sweet Marjoram; Tarragon, French Tarragon; Thyme

Soil: Well-drained, sandy, low to moderate fertility. Do not overwater.

Propagation: Start seeds early indoors, about eight to ten weeks before the frost-free date. Plant several seeds in a pot. Transplant this slow-growing herb to the garden once all danger of frost has passed. Plant a second crop in early July for a steady supply during the season.

Diseases: Generally disease- and pest-free but can develop fungal root rots or leaf spot diseases in wet soils or high-humidity conditions. Plant in containers or raised beds if this is a problem.

Spacing: Space plants 6 to 8 inches apart.

Harvesting Guidelines: Snip off stem tips as needed. When harvest-ing for drying, cut whole stems a few inches above the soil surface. Pick only when plants are dry, and harvest for preserving before plants flower.

Preserving Techniques: Hang upside down to dry, or strip stems and dry in an electric dehydrator. Freeze leaves in ice cubes or minced in oil or butter.

Exposure: Full sun but will tolerate filtered shade.

Soil: Well-drained, fertile, and rich in humus. Add ½ inch compost to the growing area each spring.

Propagation: Tarragon does not come true from seed; buy plants and propagate them by stem cut-tings in the fall. Root and overwinter the stem cuttings indoors and plant out the following spring. Divide three- to four-year-old plants in early spring. Plants die back in fall and appear again in early spring.

Spacing: Allow 2 feet in all direc-tions.

Harvesting Guidelines: Once plants are 8 to 10 inches tall, cut stem tips in the morning after the dew has dried. When harvesting for drying or preserving, cut stems to within a few inches of the soil sur-face in late June. Stems may also be cut back in fall before frost.

Preserving Techniques: Dry in an electric dehydrator to keep fla-vor and color. Freeze for the best results, either in ice cubes or a butter or oil mixture. Tarragon pre-served in white wine vinegar can be used in dishes where the vinegar is an asset.

Exposure: Full sun but tolerates partial shade.

Soil: Well-drained, sandy soil with low to moderate fertility. Do not overfertilize or overwater.

Propagation: Some cultivars can be grown from seed but others can-not. If planting seeds, start them indoors about 8 to 10 weeks before the last expected frost. Plant five to eight seeds to a pot or cell. Trans-plant to the garden a week before or after the frost-free date. Stems can be layered in late spring to early summer. Take cuttings in spring and divide plants in spring or fall. To have a steady supply of non-flowering, harvestable plants all through the season, grow a few plants each of 'Silver', 'Lemon', 'English', and 'German' thyme.

Spacing: Allow 1 foot between plants.

Harvesting Guidelines: Harvest before or after plants are in bloom. Cut stem tips or whole stems before the plant blooms, leaving at least 2 inches of stem above the soil sur-face. After bloom, take only stem tips to let the plant build resources for the coming winter.

Preserving Techniques: Bunch stems with a rubber band, and cover with a paper bag. Hang the bag upside down to air-dry. Leaves will fall into the bag as they dry.

3

Growing Fruits, Berries, and Nuts

THE FLAVOR OF HOMEGROWN FRUIT is a revelation. Most of us expect a noticeable difference in taste between a homegrown tomato and its grocery store counterpart, but we assume that fruit is fruit. However, one taste of a peach you've grown yourself will reveal what fresh really means. You'll also benefit from being able to enjoy fruit that comes from plants selected for their flavor rather than their ability to hold up to a long truck ride to the supermarket. Perhaps best of all, you'll know exactly what went into your harvest.

On a commercial scale, fruiting plants can be difficult to grow because of pest and disease problems. On a backyard scale, the major fruit pests and diseases are relatively easy to manage if you follow the guidelines in this chapter for selecting plants and their general care.

Many fruit trees can handle even the harshest winter conditions. However, rather than trust hardiness ratings, check with local gardeners to confirm which plants will thrive in your region.

Fruit Trees

Climate determines whether or not you can grow a particular fruit in your area. For example, though few Vermonters expect to be able to grow oranges or bananas in their backyards, they are usually surprised to learn that sweet cherries are also beyond their climatic reach. Similarly, it is a rare apple that can survive and produce in Zones 8b, 9, and 10. Two factors are responsible for plants' climatic preferences—their tolerance to temperature extremes and their requirements for certain numbers of "chilling" or "heating" hours. (See the box, "Heating and Chilling Hours," page 86.)

Trees and shrubs are similar to other plants in their ability to stand extreme cold or heat. A prolonged period of below-freezing temperatures is as certain to kill a lemon tree as it is to kill a tomato plant. But just as some tomato plants can stand cooler temperatures than others, certain fruit trees are bred to tolerate warmer or colder conditions where January thaws are common.

Hardiness Ratings

Fruit cultivars, specific plant varieties developed for desirable characteristics, are given hardiness ratings, just like other perennials. However, the hardiness rating doesn't tell the whole story. No matter where you live, try to buy plants that have been raised in your region. This is particularly important in Zones 3, 4, and 5a because conditions within these zones can vary due to microclimates. In the Northeast, temperatures fluctuate widely during the winter and early spring, but in the Midwest and the Prairie Provinces of Canada, they are much more stable. Because alternating thaws and freezes are much harder on plants than a period of steady cold, trees grown in the Midwest sometimes have difficulty adjusting to New England conditions. In western Massachusetts for example, January thaws are common, while they're rare in northern Michigan.

Heating and Chilling Hours

Fruiting plants require sustained cold or warm temperatures. "Heating hours" are the number of hours above 65°F to which a plant is exposed. If the heating requirements are not met, the plant will decline, and in some cases, fruit will not develop or mature.

"Chilling hours" are the number of hours below 45°F. If a plant is not exposed to an adequate number of chilling hours, it will not be able to break dormancy in the spring. Because the plant cannot leaf out, it dies. A lack of chilling hours is usually the culprit when a proven northern cultivar fails to grow in a southern location.

Various cultivars of the same species have different chilling-hour requirements. When you buy a plant, ask how many chilling hours the cultivar requires. A nursery may list them as low (300 to 400 hours), moderate (400 to 700 hours), or high (700 to 1,000 hours). To learn the average chilling hours in your area, check with your Cooperative Extension Service or local horticultural societies.

Pollinators

The flowers of most apples, pears, blueberries, and sweet cherry cultivars are "self-infertile," meaning they cannot pollinate themselves. Instead, insects must fertilize them with pollen from an entirely different cultivar. To make this even more complicated, certain cultivars can pollinate one another, but others can't.

When you buy fruit plants, check with the nursery or mail-order distributor to learn which plants and trees to pair with which, and be sure to buy those pollinators at the same time. Some growers also graft, or attach, selected scion (or top growth) branches of different cultivars to the tops of their trees. However, grafting is not always reliable, and it is a difficult process with which to obtain predictable results. For inexperienced fruit growers, it is usually recommended to begin by growing compatible trees, and only later graduate to the art of grafting trees.

Maintenance Requirements

Some tree species are harder to successfully grow than others. In Zone 5, for example, many peaches are prone to frost damage at either end of the season. If you choose to grow peaches in Zone 5 anyway, plan from the beginning to take the steps necessary to protect them from frost—draping the tree with plastic and placing jugs of hot water underneath it is one method.

Similarly, some apple cultivars are relatively pest- and disease-free while others are more susceptible. If you know the characteristics of a cultivar before you buy, you can choose to prepare for potential problems or sidestep them by selecting a more resistant plant.

Research your choices before you purchase any fruiting plant. Begin this process by searching out other fruit growers through local gardeners' associations. Growers in your own area will be able to tell you specifics about the performance of various trees in your climate.

Full-size fruit trees make a handsome, productive addition to your yard. If space doesn't allow or you simply don't need all the fruit a full-size provides, consider a dwarf or semidwarf tree.

Big or Small?

Tree sizes can vary naturally, without any help from breeders. The smallest naturally growing fruit trees, which are called "genetic dwarfs," grow only 10 to 15 feet high, making them easy to maintain. However, the flavor of their fruit usually leaves much to be desired, and they tend to be quite susceptible to diseases.

In the early 1900s, horticulturists discovered that when they grafted top growth from a standard tree (with good flavor) onto the roots of a genetic dwarf, the fruit remained the same size, but the tree grew only as large as the dwarf. In some cases, resistance to diseases increased, too. Since then, breeders have developed several different dwarf rootstocks and have worked with combinations of rootstocks and grafts.

Dwarf trees have several advantages. Because they are so small, it is easy to give them adequate attention. Their yields per tree are lower, but a well-tended orchard of dwarf trees gives as much or more fruit per square foot as an orchard of standard trees. Picking is easier and safer, too. With a small stepladder, you can reach the fruit on every tree.

But there are some disadvantages to dwarf trees. Most dwarf trees have a shorter life expectancy than standards of the same cultivar, and almost all of them are shallow rooted. Consequently, they don't survive harsh winters well. Their roots are also brittle; a strong windstorm can knock over a mature dwarf tree. To counteract this, some growers support their dwarf trees with post-and-cable trellises.

Growing a fruit tree in a container saves space, is decorative, and makes it easy to move the plant as the season warrants.

Trees in Containers

Many dwarf fruit trees grow well in containers. Because container soils dry out so quickly, look for containers with the thickest walls possible. Most dwarf trees will grow in a pot that is 2 feet wide by 3 feet deep, although more room never hurts. Because fruit-bearing trees can be top-heavy, the best containers are wide enough at the bottom that they won't tip over in the wind.

Soil drainage is essential. If you are planting in an old barrel or other container, drill several ¾-inch holes in the bottom. Add a layer of nylon window screening, and then fill the container with a nutrient-rich and quickly draining soil mix.

Protect Container-Grown Plants from Frost

Protect the roots of your trees over the winter if you live in an area where temperatures go below freezing. Sink the pots in soil during the late fall to give the best protection. If you have an unheated area, such as a garage, where temperatures remain above 32°F, move the container there once the plant is dormant. If neither of these options is possible, move the container to a somewhat protected niche, and pile plastic bags filled with autumn leaves under, over, and all around the container. Tie a strong plastic tarp around this construction to hold it in place. In the spring, wait until heavy frosts have subsided to bring the container out in the open again.

Planting Fruit Trees

Most fruiting plants require full sun to bear well. However, full sun on a southern slope is more intense than full sun on a northern one, and your plants will respond accordingly. If you live in an area with late spring frosts (which can damage the blooms of a tree), plant on a northern slope if at all possible. Because it is cooler there than on a southern slope or on level ground, trees will bloom a little later. As a result, there will be fewer years when you need to protect the blossoms from late-spring frosts.

Room to Grow

Try to plant your trees at the same depth that they were growing at the nursery. This level will be easy to determine if the tree is balled and burlapped or in a container, but even if it is bare-root, you should be able to see the soil mark on the trunk. Proper planting depth is particularly important in the case of plants with grafts.

Always plant so that a graft union is 2 to 3 inches above the soil surface. Ask the nursery how deep to plant the rootstock because it can vary depending on cultivar, graft, and rootstock.

How to Plant Fruit Trees

Make a hole with sloping sides, as deep as the depth of the root-ball and twice that width at the top. To help the roots penetrate the surrounding soil, use a hand-held claw to roughen up the sides of the hole. After the tree is in position, backfill with the soil from the hole without amending it in any way. Form a mounded ring of soil 12 to 18 inches from the trunk to create a 6-inch-deep basin for watering. Water deeply when the hole is one-half to three-quarters filled and again when the backfilling is complete.

Space

Adequate space is essential for fruiting plants. Space your trees a distance equal to their mature heights. If trees with differing mature heights are planted next to each other, use the taller one as your guide.

You can use the space between trees to grow other crops while the trees are young. Both bramble berries and strawberries are ideal to plant between young fruit trees, but you can plant anything from pumpkins to cutting flowers. Eventually, plant a carpet of small flowering plants to feed the many beneficial insects that can help keep fruit insect pests in control.

Soil Characteristics

Look for soil that has good drainage, a pH of 6.0 to 6.8, and moderate to high fertility levels with a high concentration of organic matter. Because so few garden soils are totally ideal, you'll probably have to do some soil improvement. (See pages 40–42.) Only two circumstances should prevent you from planting fruit in an area: one is a pH that is more than one point away from ideal, and the other is extremely poor drainage.

Air Drainage

Many fungal diseases become more troublesome on plants growing where the air is stagnant. Fruit trees are especially susceptible to fungi, so try to site them where the prevailing breeze will rustle through their leaves each day.

Look for the soil mark on the trunk of the tree as a guide to planting at the right depth.

Pruning Fruit Trees

Good pruning strengthens the tree and makes it more productive. Pruning when the tree is young gives it a strong framework of scaffold branches that are positioned to allow both light and air into the center of the tree. As the tree matures, pruning maintains a balance between shoot and fruit production. It also removes weak, damaged, or diseased growth. Finally, it allows air and light to reach all parts of the tree, stimulating new growth where you want it.

Tree Anatomy

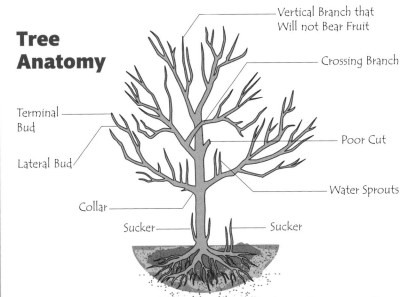

Vertical Branch that Will not Bear Fruit

Crossing Branch

Terminal Bud

Poor Cut

Lateral Bud

Water Sprouts

Collar

Sucker

Sucker

Pruning is much easier once you can identify the various parts of a tree. Though each tree is different, you'll see many of these structures on your own tree. Prune to bring out the best characteristics of each of your trees and minimize their weak points.

The Purpose of Pruning

Pruning is a mystery to most of us. We're afraid we'll cut off the wrong thing and do the tree irreparable damage. A few principles of plant biology are a helpful guide to how pruning shapes plant growth.

Fundamental to the process are *auxins*, the hormones that regulate plant growth. They promote growth at the tip of the stem but inhibit the development of buds lower down on that stem. This effect, known as *apical dominance*, is what prevents a tree from growing a new limb at every bud. When you prune off the tip of a stem, you remove the site where the auxin is produced, stimulating dormant buds lower in the plant. By choosing how much of a branch to remove, you can direct the plant's response. For example, pinching off the very tip of the branch encourages branching just below the pinched area. Shortening a stem by a third stimulates the growth of buds just below the cut. Pruning two thirds of a stem promotes even more vigorous growth of a bud very low on the branch. In general, the buds nearest the cut make the most vigorous upright growth, while those lower down

on the stem grow at wider angles. The bud that develops nearest the tip of the stem will eventually become a dominant branch and produce auxins to inhibit the growth of buds below it. All pruning cuts used to cut back stems are referred to as *heading* cuts.

Thinning cuts are those that totally remove growth. Use thinning cuts to prune off weak or poorly positioned branches, water sprouts, and sprouts from the rootstock. When you remove a branch along a stem, the tip of the stem grows even more vigorously and latent buds are less likely to develop. If you want the stem to form a branch in a different spot than the branch that you pruned, remove the site of auxin production by heading back the stem an appropriate distance.

Thinning, or pruning off, the developing fruit not only evens out production from one year to the next but also increases the size of the remaining fruit. It is best to thin the fruit twice, once just after the blossoms have dropped and again just after the June fruit drop. As a general rule, allow a space two to three times the size of the mature fruit between the fruit you leave on the tree.

When to Prune

Most heading-cut pruning on apples and pears is done in late winter or very early spring, while the plant is still dormant. When the tree comes out of dormancy, the lack of auxin production at the headed-back stem tips causes lateral buds to grow vigorously.

In contrast, very little to no dormant pruning is done on peaches and nectarines because it promotes early blooming. The first defense against frost damage to these trees is to wait until the tree is already in bloom to make heading cuts. If done at this time, the tree still responds by producing new growth along branches that have been headed back.

Pruning in summer reduces rather than stimulates regrowth. You can use this to your advantage when

Encouraging Branching

Sometimes, despite the best possible pruning cuts, a bud stubbornly refuses to develop. If you really need that branch to balance the tree, you can often stimulate growth by making a small notch in the stem about an inch above the bud.

you are making thinning cuts to bring light into the center of the tree or to remove unwanted growth. Heading cuts done in midsummer can have the effect of stimulating the formation of flower buds rather than those that will grow into shoots.

Tree Shapes

The shape of your tree will be determined in large part by the type of tree it is. In general, dwarf and semidwarf trees are best pruned as vase or modified central-leader trees, while standard trees can be pruned as central-leader or modified central-leader trees.

CENTRAL-LEADER PRUNING depends on well-placed scaffold branches to allow light into the center.

MODIFIED CENTRAL-LEADER PRUNING requires you to cut out the central leader after the fourth or fifth scaffold branch grows.

VASE, OR OPEN, PRUNING is done by cutting back the leader to form a completely open center.

Before and After Pruning

BEFORE PRUNING: The branches of this tree are too closely spaced and prevent light from reaching the center of the tree.

AFTER PRUNING: By eliminating weak or misplaced branches, you open up the center of the tree so light can reach branches uniformly.

First-Year Pruning

Begin by learning about the natural growth habits of your tree and the shape that most growers give it. Common shapes—central leader, modified central leader, and vase, or open, shapes—are illustrated opposite.

If you buy a tree that already has some branches, prune to retain all well-positioned branches. Pruning, even if done when the tree is dormant, stimulates growth near the cuts but sets back the tree's overall growth. Ideally, the bottom branch will be about 2 feet above the soil surface and will be growing at an angle of at least 40 degrees from the trunk. If as many as two other branches are spaced 6 to 8 inches apart on the trunk, grow or can be trained to grow at a 40-degree or greater angle, and are positioned in a neat spiral around the trunk, you'll want to retain them, too. Head back each of the retained branches to a few inches, always cutting just above an outward-facing bud—unless the tree is a dwarf that you are growing in a trellised system. Thin the branches that are growing too closely to a desired branch, that do not help to form a spiral around the trunk, or that grow at such a strong angle that they cannot be trained to a more horizontal position. Head back the central leader to about 3 feet from the soil surface.

Second-Year Pruning

In the second year, pruning is dictated by the tree's form. If you are growing a central-leader tree, head back the leader again so that a new tier of *scaffold* branches will develop—the primary limbs that radiate from the trunk of the tree. Also thin out undesirable growth, and head back the branches. Cut back about a third of the previous year's growth from the central leader.

With a modified central-leader tree, cut back the central leader to just above the fourth scaffold branch. If the tree lacks this branch, head back the central leader. By the third season, there will be enough branches for you to cut back the central leader. Head back the branches you have decided to keep, and then thin out any undesirable growth. With trees with open centers, the central leader is usually removed during the second year. By this time, the trees usually have grown the three branches that form the vase shape. Thin undesirable growth.

To widen a branch angle, insert a wooden toothpick between a developing branch and the leader to force the branch into a more horizontal form. Weighing the branch with wooden clothespins or tying it down also gets the job done.

Pruning Guidelines

- Prune only as much as necessary to create the desired shape, to allow light into the center, and to keep the tree healthy and bearing well.
- When pruning a dormant tree, wait to prune until the wood has thawed in the morning sun.
- Use only appropriate tools, and sharpen them before every use. Sterilize tools with a 10 percent laundry-bleach solution after every cut on a diseased tree.
- When pruning off diseased or insect-infested growth, cut back the branch at least 6 inches beyond the site of the problem.
- Remove and destroy all diseased and insect-infested wood. Do not leave it near the tree or compost it.
- Cut all branches just below the collar. (See illustration on page 89.) When pruning branches back, make the cut above an outward-facing bud.
- Do not use tree paint to protect wounds.
- Make clean cuts using sharp tools.
- Prune peaches and nectarines only when they are actively growing, not when they are dormant. With your finger, rub off undesirable buds during the growing season so that you won't have to prune them off the following year.
- Prune off diseased growth immediately. If you wait until the tree is dormant, the problem can worsen.

Angle all cuts so that excess moisture drips off, discouraging organisms that thrive in high moisture conditions.

Use a small pruning saw to cut branches that are thicker than ½ in.

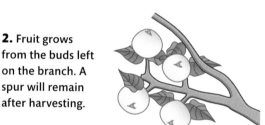

1. Spurs form best on lateral branches pruned back to 3 or 4 buds.

2. Fruit grows from the buds left on the branch. A spur will remain after harvesting.

3. Additional spurs will form each year. They will be just below the fruits that grow.

4. After a few years, spurs will become too crowded. Prune out older ones.

5. Thin the spurs as necessary in late winter or early spring every year.

Caring for Fruit Trees

The first three to five years shape the future of a fruit tree. Keep on top of pruning, watering, mulching, managing pests and diseases, and winter preparation. Give the tree a good start, and your chores will ease up as the tree matures.

Trees also need adequate water supplies for a good start. Water the planting hole deeply when you plant. Unless you live in a dry climate or your soil is unusually dry, do not water again until you see new growth. Soggy soils can invite root rot. But once the buds begin to plump out, put the tree on a regular watering schedule.

Set up a rain gauge near the tree. Your tree will require between 1 to 2 inches a week. If rainfall is less than this, you will need to irrigate. See pages 96 to 100 for the required quantities of water.

When you plant, make a 3- to 4-inch mound around the tree about 1 to 1½ feet from the trunk to form a basin. Soak this area thoroughly at an equivalent of at least 2 gallons per square foot of root area. As the tree grows each year, expand the area of the basin so that it extends just beyond the canopy tips. When

Fertilizing

In good soil, an application of 5 to 10 pounds of compost per tree every spring maintains an adequate nutrient supply. However, because this application simply replaces the nutrients that were lost during the year, you may need to supplement with greater quantities and other nutrient sources.

Test the leaves for nutrient content every two to three years. Laboratories that provide this service can be found on the Internet by searching under "plant and tissue testing." Using the recommendations of the lab, apply needed nutrients in slow-release forms to prevent burning the roots. For example, alfalfa or soybean meal will add nitrogen; rock phosphate and kelp supply phosphorus; and greensand provides potassium.

you calculate watering requirements, remember that the roots actually extend into the soil about one and one-half times the diameter of the top growth.

Well-tended fruit trees, left, can be as decorative as they are productive.

A rain gauge, above, is a sure way to monitor how much water your fruit trees are getting and how much you need to add by way of irrigation.

93

3 Growing Fruit, Berries, and Nuts

Mulching

Using organic mulch around your fruit trees supplies food for nutrient-supplying microorganisms, holds soil moisture, and prevents weed growth. But it can also provide a cozy habitat for rodents and other tree pests.

When you plant, apply a layer of mulch outside the mound area. This placement guarantees at least a foot of bare soil will surround the trunk. Although this distance may not seem like much, most mice will think twice about nibbling on something that far from cover.

Apply a deep enough layer of mulch to ensure that it will keep weeds down and slow evaporation from the soil. Depending on your climate and the biological activity in your soil, up to 6 inches of straw or 2 inches of shredded bark mulch will disappear over the season. Replace it as it degrades.

Dropped fruit is common at harvest time, top right. Clean it up before winter sets in. Left on the ground, it will harbor disease and insects.

Before winter sets in, remove all old mulch from around the tree, right, and add it to a hot compost pile. Debris left around the tree can harbor disease and pests.

Espaliered Fruit Tree

An espaliered fruit tree takes some work but is a real space-saver that can make an underutilized corner of your farmstead productive. The most appropriate trees for espaliers are dwarf or semidwarf cultivars of apples, pears, plums, cherries, apricots, and nectarines. To create an espalier, plant the tree, and allow three or four branches to grow along the horizontal supports, tying them into place as they lengthen. Small shoots will grow from these branches during the summer. In the late summer or early fall, after the bark on these shoots has turned brown and the bases of the shoots have become woody, cut them back to retain three leaves beyond the basal cluster of leaves. When the tree is dormant, prune the fruiting spurs so that the fruit will be well spaced.

An espaliered fruit tree trained to grow on a wall saves space and can shield the tree from prevailing winds.

Sanitation

Good sanitation helps to keep pests and diseases from building up significant populations near your garden plants. But nowhere is it more important than around fruiting plants. Don't leave pruning waste on the ground. Pests and diseases overwinter in the bark. If you have a chipper, put the pruned branches through it before adding them to a hot compost pile. If you don't have a chipper, dispose of the branches far away from the trees, or burn them.

Small fruits may drop to the ground in early June if they are infested with plum cucurlio or codling moth larvae. Interrupt the pest life cycle by picking up the fruit and destroying it. Do not put June drops in a compost pile or bury them. Instead, put them in a plastic bag destined for the landfill, or burn them.

At the end of season, when you are preparing your trees for winter, remember to rake off all the old mulch around the trees along with the fallen leaves. Use this material to build or add to a hot fall compost pile. If your supply of nitrogen-rich green material is low, add alfalfa pellets to boost the nitrogen level.

Painting fruit trees with latex or wrapping them with a tree guard protects them from winter sunscald.

Monitoring for Pests and Diseases

Pests and diseases take hold so slowly you'll be hard put to notice them until they've done real damage. Once a week, take the time to examine your plants. Check the top and bottom surfaces of the leaves; examine the bark; and check all new growth to be certain it is clean and healthy. After you have grown each type of fruit plant for a while, you will know which pests and diseases are most likely to strike and how to manage them so that they inflict the least possible damage.

Preparing for Winter

Fall is one of the busiest seasons for the fruit grower. Use this time to be sure your trees go into the winter with the best possible sanitation. In addition to raking up old mulch and debris, remove all the shriveled fruit that is still on the tree and destroy it. Add new mulch around the tree, 12 inches from the trunk. Protect the trunk from winter sunscald by wrapping it with a tree guard or painting it with white latex. Enclose painted trees with fine-mesh wire to keep rodents away.

Fruit Trees, Plant by Plant

APPLE

Delicious, bountiful, and readily preserved, the apple has been a farmstead staple for centuries. Whether made into sauce, canned, dried, or stored raw, apples are a great food source. And don't forget cider—the real reason Johnny Appleseed was so welcomed on the frontier. Young spur-bearing apple trees grow about 6 to 12 inches each year, while non-spur-bearing trees grow anywhere from 1 to 2 feet. Once trees are mature, growth slows to about 6 to 10 inches a year for both types.

APPLE

Zones: 3 to 9.

Size: 8 to 30 feet tall, 8 to 40 feet wide.

PLANTING

Exposure: Full sun, preferably on a north-facing slope.

Soil: Well-drained, moderately fertile, pH of 6.5 to 6.8. In early spring each year, apply 5 to 10 pounds of compost to the soil surface, from about 6 inches from the trunk to just beyond the drip line.

Spacing: Plant trees as far apart from each other as they will grow tall.

Pollinators: Check with the nursery or supplier to learn which of the cultivars that bloom at the same time are compatible. Try to buy pollinating cultivars at the same time to be sure to get them.

CULTURAL CARE

Watering: Apply approximately 2 gallons per square foot of root area every week.

Mulching: After planting, leave an 8- to 12-inch-diameter area around the trunk weed-free and clear of mulch. Beyond this, mulch with 4 to 6 inches of straw or 3 to 4 inches of a denser material. In a lawn, mulch a minimum of 4 square feet around each tree.

PRUNING

After Planting: Choose future scaffold branches and thin the others. Head back the leader and branch tips.

Routine Pruning: Prune to modified central-leader, central-leader, or open-center form, depending on size of tree. Make heading cuts while the tree is dormant, and thin as much as possible in summer.

Pruning Mature Plants: Thin spurs and fruits as necessary, prune out damaged or infested growth, and thin to allow light and air into the center of the tree.

HARVESTING

Fruiting: Depending on the cultivar, plants fruit on 1- to 10-year-old wood, on spurs or tips of branches.

Years to Bearing: 3 to 8, depending on rootstock and cultivar.

Harvest Season: Late summer to late fall.

Yields: 10 to 30 bushels for standards and semidwarfs, less for dwarfs.

EUROPEAN APRICOT, MANCHURIAN APRICOT, SIBERIAN APRICOT

Zones: European, 5 to 9; Manchurian and Siberian, 3 to 8.

Size: 4 to 25 feet tall, 6 to 25 feet wide.

PLANTING

Exposure: Full sun in an area protected from winds and early and late frosts.

Soil: Well-drained, sandy loam soils, moderate to high fertility.

Spacing: As far away from other trees as the tree's mature height. Apricots make ideal espalier or cordon trees.

Pollinators: Apricots, even the self-fruitful European cultivars, bear best with pollinators. Buy several compatible trees. 'Nanking' cherry is a good pollinator.

CULTURAL CARE

Watering: At least 2 gallons per square foot of root area every week or 1 inch of rainfall.

Mulching: After planting, leave an 8- to 12-inch-diameter area around the trunk weed-free and clear of mulch. Beyond this, mulch at least a 4-foot-square area with 4 to 6

Apple; European Apricot, Manchurian Apricot, Siberian Apricot; Sweet Cherry

inches of straw or 3 to 4 inches of a denser material.

PRUNING

After Planting: Retain any branches that will make good scaffold limbs, and head them back unless you are training the tree as an espalier.

Routine Pruning: Prune to a modified central-leader or open-center form. Head back branches in spring, and make as many thinning cuts as possible in summer. Thin fruit to at least 2 inches apart.

Pruning Mature Plants: Stimulate new growth by thinning small branches after they are 3 years old and keeping all branch tips headed back.

HARVESTING

Fruiting: Apricots fruit on 1-year-old spurs and branch tips.

Years to Bearing: 3 to 5.

Harvest Season: July and August. Fruit will not ripen after being harvested, so pick when it has softened, but before it drops.

Yields: 1 to 2 bushels for dwarfs, 3 to 4 bushels for standard-size European cultivars.

APRICOT

Tasty eaten fresh, the apricot is also a versatile cooking ingredient and popular dried fruit, and it makes excellent jam.

SWEET CHERRY

Though best eaten fresh, sweet cherries make a refreshing juice and a pleasant flavoring for tea. Some cooks make pies with sweet cherries to eliminate sugar in the recipe.

SWEET CHERRY

Zones: 5 to 9.

Size: 6 to 35 feet tall, 8 to 40 feet wide, depending on rootstock and cultivar.

PLANTING

Exposure: Full sun.

Soil: Well-drained, deep, and moderately fertile, pH of 6.0 to 6.8. Maintain fertility with 5 to 10 pounds of compost around each tree in very early spring.

Spacing: As far apart as the mature tree will grow.

Pollinators: Almost all sweet cherries require pollinators, and the few that are self-fruitful yield better if they have a pollinator. Not all sweet cherries are compatible.

CULTURAL CARE

Watering: A minimum of an inch of rainfall or 2 gallons per square foot of root area every week. Keep soil moisture consistent when fruit is ripening. Overwatering promotes cracking.

Mulching: Leave between 8 and 12 inches of bare soil around the trunk. Beyond that, apply 3 to 4 inches of dense mulch, such as shredded hardwood bark, or 6 to 8 inches of a loose mulch, such as straw.

PRUNING

Fruiting: Fruit forms on 1- to 10-year-old spurs.

After Planting: Choose which scaffold limbs you will retain, and thin the other branches out. Head back the retained limbs to stimulate branching.

Routine Pruning: Prune and train to an open-center form whenever possible. Keep up with training branches to spread because the natural growth habit is quite upright. Remove scaffold branches that form opposite each other because their weight may split the trunk. Do not thin fruit.

Pruning Mature Plants: Cherries need minimal annual pruning. Remove all diseased or damaged growth, and thin out branches that prevent light from entering the center of the tree.

HARVESTING

Years to Bearing: 4 to 5.

Harvest Season: July to August.

Yields: ¾ bushel for dwarfs, 1 bushel for semidwarfs, 2 bushels for standards. Sweet cherries require a minimum of 1,000 chilling hours but will not fruit well where summer temperatures regularly exceed 90°F. If fruit appears hard, shriveled, and blotchy, have the leaves tissue-tested.

SOUR CHERRY

Typically the stuff cherry pies are made of, sour cherries enhance a wide range of desserts. Sour cherries make a tasty jam and freeze well for future use. Sour cherries can be trained as espaliers. If you live in the North, choose a sour cherry such as 'Evans', which is known for its hardiness.

SOUR CHERRY

Zones: 4 to 8.

Size: 8 to 20 feet tall, 10 to 25 feet wide.

PLANTING

Exposure: Full sun, but trees can tolerate a bit of light shade in the afternoon. In Zones 4 and 5, a protected niche on a north-facing slope is ideal.

Soil: Well-drained with moderate fertility levels and a pH of 6.0 to 6.8. Maintain fertility by applying 5 to 10 pounds of compost around each tree in early spring.

Spacing: As far apart as mature tree height.

Pollinators: None required.

CULTURAL CARE

Watering: A minimum of 1 inch of rainfall or 2 gallons per square foot of root area every week. Keep soil moisture consistent when fruit is ripening. To prevent cracking, do not overwater.

Mulching: Leave 8 to 12 inches of bare soil around the trunk. Beyond that, apply 3 to 4 inches of dense mulch, such as shredded hardwood bark, or 6 to 8 inches of a loose mulch such as straw.

PRUNING

After Planting: Maintain well-positioned scaffold branches, and thin others. Head back leader and branches.

Routine Pruning: Prune to an open-center form or a modified central leader. Thin branches to allow light and air into the center of the tree. Train branches to form wide angles.

Pruning Mature Plants: Thin poorly positioned or diseased branches.

HARVESTING

Fruiting: Fruit forms on 1- to 10-year-old spurs.

Years to Bearing: 4 to 5.

Harvest Season: Mid-July to early August.

Yields: 1 bushel, dwarf trees; 2 bushels standard trees.

LEMON

If you are lucky enough to live where citrus trees flourish, you'll enjoy several months of fresh fruit each year.

GRAPEFRUIT, LEMON, LIME, ORANGE

Zones: 9 to 10. Citrus must almost always be purchased locally because rootstocks are chosen to suit particular locations. In the North, you may want to grow containerized dwarf cultivars. If so, keep the tree in a protected spot during the summer, and move it indoors long before frost.

Size: 5 to 35 feet tall, 8 to 35 feet wide, depending on species and type.

PLANTING

Exposure: Full sun, although some limes and lemons will tolerate after-noon shade in extremely hot locations.

Soil: Well-drained, high fertility and good moisture-holding capacity with a pH of 6.0 to 6.5. Nitrogen demands are relatively high, so it may be necessary to apply a cup of alfalfa pellets under compost applications. Unless the soil is unusually fertile, spread 10 to 20 pounds of compost under the tree every six weeks to two months.

Spacing: As far apart as mature trees are tall.

Pollinators: None required.

Sour Cherry; Grapefruit, Lemon, Lime, Orange; Peach, Nectarine

CULTURAL CARE

Watering: Generally about 1½ to 2 inches a week or about 3 gallons per square foot of root area.

Mulching: Use a nitrogen-rich mulch such as grass clippings or alfalfa hay to add nitrogen to the soil. If weeds become a problem, apply several sheets of newspaper under the mulch.

PRUNING

After Planting: Head back branches.

Routine Pruning: Citrus trees do not require specialized pruning because they naturally form the most productive forms. After harvesting, prune to remove dead, weak, or poorly positioned branches. On oranges, remove all suckers. On lime bushes, thin old stalks periodically.

Pruning Mature Plants: Thin to allow light and air into the center of the plant.

HARVESTING

Fruiting: Plants bear on the current year's growth.

Years to Bearing: 2 to 4.

Harvest Season: Varies with fruit type and location. Check with your supplier.

Yields: Varies depending on fruit type and tree size.

PEACH

Beyond pies and cobbler, nectarines and peaches can be made into butter, jellies, jams, and preserves, and they freeze well. Some outdoor cooks love them grilled.

PEACH, NECTARINE

Zones: 5 to 9.

Size: 4 to 20 feet tall, 6 to 25 feet wide.

PLANTING

Exposure: Full sun in a protected area on a north-facing slope.

Soil: High fertility, sandy, and well-drained with a pH of 6.5 to 6.8.

After petal drop, supplement fertility by spreading 1 cup of alfalfa meal, 1 cup of gypsum, and 1 cup of rock phosphate over the root area before applying 10 to 20 pounds of fully finished compost.

Spacing: As far apart as the mature height of trees.

Pollinators: None required for most cultivars.

CULTURAL CARE

Watering: Trees require a minimum of an inch of moisture a week. Supplement rain with irrigation if necessary, supplying 2 gallons per square foot of root area.

Mulching: To discourage rodents, keep 8 to 12 inches of soil around the trunk free of mulch and vegetation. Mulch beyond this with 6 to 12 inches of straw or 3 to 6 inches of shredded hardwood bark. In the North, apply at least 1 foot of new mulch before snowfall to protect the roots over the winter.

PRUNING

After Planting: Keep well-positioned scaffold branches, and thin the others. Head back the leader and branches to stimulate branching.

Routine Pruning: Prune to an open-center or modified central-leader form. Nectarines require a great deal of pruning to remain productive and healthy. Wait to prune until the tree is in bloom. Each year, cut back the longest branches, including the leader.

Pruning Mature Plants: Continue to cut back branches and the leader. Thin fruit to 6 to 8 inches apart. Thin weak, dead, and crossing branches. Thin to allow light and air into the center of the tree.

HARVESTING

Fruiting: Fruits form on 1-year-old wood.

Years to Bearing: 3 to 5.

Harvest Season: Late July, August, and early September, depending on cultivar. Pick early if frost threatens because fruit continues to ripen off the tree.

Yields: 3 to 5 bushels for standard trees, 1 to 3 bushels for dwarfs.

3 Growing Fruit, Berries, and Nuts

Delicious fresh, pears also make tasty jams, butter, and sauces. Canned pear pieces extend the joy.

PEAR, ASIAN PEAR

Zones: 4 to 9.
Size: 8 to 40 feet tall, 10 to 25 feet wide, depending on cultivar.

PLANTING

Exposure: Full sun, high circulation.
Soil: Moderate fertility and well-drained soil. Pears can tolerate clay soils if humus content is high to promote drainage and pH is between 6.0 and 6.5. Do not overfertilize.

Spacing: Space as far apart as the mature trees are tall.
Pollinators: Not all pears are compatible, check with your supplier.

CULTURAL CARE

Watering: Trees require 1 inch to 1½ inches of moisture a week. Apply 3 gallons of water per square foot of root area.
Mulching: Keep 1 foot of bare soil around the trunk. Beyond this area, mulch with 6 inches of straw or 3 to 4 inches of a denser material.

PRUNING

After Planting: Choose well-positioned branches to keep as scaffold limbs, and thin poorly positioned growth. Head back the leader and branch tips.
Routine Pruning: Prune and train to a central leader or modified central leader. Because fire blight infections may require pruning off an entire branch, retain more scaffold branches than you would if it were an apple tree.
Pruning Mature Plants: Remove growth that prevents air and light from getting into the center of the tree. Do not thin fruit unless set is unusually heavy. Thin old wood when the spurs are 7 or 8 years old.

HARVESTING

Fruiting: Fruit forms on 1- to 10-year-old spurs and the tips of lateral branches.
Years to Bearing: 4 to 7.
Harvest Season: August to October, depending on cultivar. For the best texture and flavor, pick before the fruit is fully ripe, and ripen in a root cellar or refrigerator.
Yields: ½ to 1½ bushels for dwarfs, 2 to 4 bushels for standards.

When dried, plums become prunes. When canned, they are often blanched first to remove the skin.

EUROPEAN PLUM, JAPANESE PLUM

Zones: European, 4 to 9; Japanese, 6 to 10.
Size: European, 1 to 20 feet tall, 10 to 20 feet wide; Japanese, 10 to 20 feet tall, 10 to 20 feet wide.

PLANTING

Exposure: Full sun.

Soil: Well-drained, high-fertility. European plums tolerate heavy clay; Japanese plums prefer sandy soil.
Spacing: As far apart as mature trees are tall.
Pollinators: Check with your nursery for the best pollinators.

CULTURAL CARE

Watering: Plants require 1 to 1½ inches per week.
Mulching: Leave 8 to 12 inches of bare soil around the trunk, and mulch beyond this area.

PRUNING

After Planting: Choose branches to retain for scaffold limbs, and thin all poorly positioned branches. Head back the leader and all retained branches.
Routine Pruning: Prune and train European plums to a modified central-leader form and Japanese plums to an open-center form.
Pruning Mature Plants: European plums require very little pruning. The growth on Japanese plums requires thinning. Allow 4 to 6 inches between developing fruit of Japanese plums, and thin European plums to two per spur.

HARVESTING

Fruiting: European plums fruit on 2- to 6-year-old spurs and Japanese plums on 1-year-old and older spurs.
Years to Bearing: 3 to 4.
Harvest Season: July to September.
Yields: 1 to 2 bushels for standards and ½ to 1 bushel for dwarfs.

Choosing Fruiting Shrubs

Almost all fruiting shrubs require full sun, although currants and gooseberries (*Ribes* species) can tolerate light afternoon shade if they are growing in a hot area. Good air circulation is imperative to prevent fungal diseases. Whenever possible, site fruiting shrubs where the prevailing breezes will blow through them to prevent moist air from settling around the leaves. Locate early-blooming plants, such as currants and gooseberries, on a north-facing slope to delay their bloom as long as possible, or place them in a protected spot where late frost is rare.

Elderberry shrubs will not only feed you, they'll provide a decorative addition to your farmscape—a good candidate for the front yard.

Supplying Nutrients

Fruiting plants of all types require soils that are at least moderately fertile. As a general rule, apply 1 to 1½ inches of compost around the base of your shrubs every spring, shaking it down into the center of the stems if necessary and extending it several feet in all directions. Apply moisture-conserving mulches over the compost. If rodents are a problem, use fine-meshed screening around the bush to keep them from nibbling the bark. Bury the screening in a 6-inch-deep trench to prevent their burrowing under it.

Add specific materials such as rock powders, kelp, and alfalfa pellets to supply additional nutrients. If you suspect nutrient deficiencies once the shrubs are planted and growing, send leaf samples to a lab that conducts tissue analysis. Use the table on page 103 to choose an appropriate nutrient source.

Because it contains relatively low amounts of immediately available nutrients, compost is suitable for mulching around your plants any time during the growing season. However, you'll need to take care with high-nitrogen materials, such as fish emulsion, alfalfa pellets, and soybean meal. If you add high-nitrogen fertilizers in late summer, plants may put out a burst of new growth that is vulnerable to winterkill. Wait until the danger of frost has passed to add a high-nitrogen fertilizer. Refrain from using such a fertilizer after mid-July.

Slow-release, natural sources of nutrients, such as potassium, calcium, and phosphorous, are generally added a year before planting a perennial crop such as a fruiting shrub. If you need to supplement after the plant is growing, apply nutrients in spring, and cover them with compost. Microorganisms in the compost will help to make the nutrients in these materials available to the plants, and rainwater will wash the nutrients deep into the root zone. Add lime in the fall before planting the winter cover crop or applying a protective layer of mulch.

Use foliar feeds in the early morning or on an overcast day to protect the leaves from burning under the droplets in bright light. Liquid seaweed, powdered kelp, and nettle and compost teas all make excellent foliar feeds. Both liquid seaweed and kelp minimize frost damage in the spring if they have been sprayed on the plant a day or two before the frost.

Growing Fruit, Berries, and Nuts

Soil

Blueberries and cranberries differ from the majority of plants in their pH requirements. While a pH of 6.5 to 6.8 suits almost all the other plants you'll grow, these plants suffer in a pH higher than 4.5. Some Northeastern soils are naturally this acid, as shown by the numbers of wild blueberries growing on hills and at the edges of northern woodlands, but most garden soils are not.

When you have your soil tested, tell the lab that you will be growing blueberries and request a recommendation for sulfur, rather than another acidifying material.

Till the soil, and apply the recommended amount of sulfur, working it into the top few inches. Mulch over this with pine needles or crushed pine bark, replacing it as necessary to keep weeds from taking over the area. It may take as long as a year for the soil to become properly acidified—keep testing with a pH meter or litmus strips. In a year, the soil should be acid enough to plant.

In addition to increased acidity, cranberries require extremely moist soil. Site them where the water table is naturally high, and plan to keep them watered well at all times. You don't have to flood them as commercial growers do as long as you maintain high moisture levels.

Growing Blueberries

Blueberries require such high acidity that you might not be able to grow them in your native soil. Instead, build boxes at least 3 x 3 x 3 feet; provide them with good drainage holes; and fill them with specially prepared high-acid soil. Mulch with pine needles, and check and adjust the pH each year.

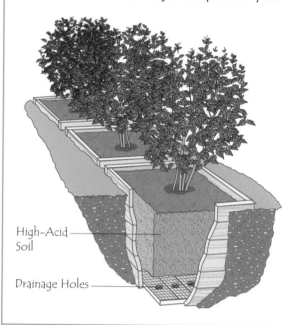

High-Acid Soil

Drainage Holes

Routine Care and Special Needs

Fruiting shrubs tend to be easier to care for than fruit trees. In most cases, the pest pressure isn't as high, and it is easier to keep them disease free. Pruning needs are lower, and the plants are more forgiving. But you can't just plant them and walk away.

Birds are likely to be the biggest problems for fruiting shrubs. Many people frighten birds away by painting balloons to look like giant eyes and suspending them from high stakes near the bushes. Others twist red and silver plastic tape through the bushes. Done well, the breeze will move the tape so that it looks like fire to a bird's eye. Some people report good results by hanging disposable aluminum pie pans in the bushes, and others

say that birds will avoid the area if you set up perches that are covered with a sticky paste designed for this purpose. The birds will be able to escape the stuff but will develop an aversion to the area.

These techniques have all been successful for many gardeners, but none of them is as foolproof as the traditional recourse—netting. If birds are taking too much fruit, build an enclosure of 1x3 or 2x4 boards, and cover all sides and the top with bird netting. If the enclosure is at least 6 inches taller than the bushes, the birds will not be able to get to the fruit.

Adequate soil moisture is crucial to fruit quality. Set up a rain gauge near the bushes, and monitor it.

Pruning Fruiting Shrubs

Bushes grow by developing new stems from the crown of the plant every year. Individual branches are not long lived, but the annual new growth means that the bush itself may be. Your job is to remove old or diseased growth to make way for healthy new stems. You'll also want to open up the center of the plant to light and air.

After planting, there are two ways to handle a fruiting shrub. In the case of blueberries, cranberries, and serviceberries, treat them as bushes and do as little pruning as possible. Thin out weak branches, and head back any damaged wood, always cutting to an outside bud. Pinch off the flowers so that the plant can put all of its strength into root and shoot growth.

Prune shrubs when they are dormant. After the first year, thin out weak or damaged growth, but wait to do major pruning until the third or fourth year.

Highbush and half-high blueberries fruit on stems 1 to 4 years old. After that, productivity declines. Thin all 5-year-old stems, cutting at the soil level. To keep the plant sturdy, head back all drooping stems and those that are less than ¼ inch thick. Always cut just above an outside bud or branch. To enlarge berry size, head the stems back so that they don't carry more than five fruit buds on each fruiting stem.

Rabbiteye blueberries are pruned in a similar fashion except that you probably won't need to head back fruiting stems because these plants are unlikely to have too many fruit buds on a stem. Also, rabbiteye plants are generally more vigorous and robust than the others.

Lowbush blueberries form new shoots from underground *rhizomes*—the main horizontal roots of the plant—as well as from buds on the branches. The biggest and best fruits come from shoots growing from the rhizomes. Prune to stimulate this kind of fruiting. The third year after planting, cut half of your plants to the ground while they are dormant. The half you don't prune will set fruit that year. During the next dormant period, cut all the stems to the ground on the half that you allowed to fruit that year. Alternate pruning in this way to keep your bushes healthy and bearing well.

Common Fruiting Shrubs

Fruiting shrubs are likely to be the hardest-working plants on your farmstead. Some fruiting shrubs are pretty enough to be decorative, while others make good hedges.

NAME	ZONES	SOIL NEEDS	PRUNING NEEDS	FRUITING TIME	POLLINATORS REQUIRED
Blueberry, lowbush	3 to 8	Acid	Light	Late summer/fall	Yes
Blueberry, rabbiteye	6 to 9	Acid	Light	Late summer/fall	Yes
Blueberry, highbush	4 to 8	Acid	Light	Late summer/fall	Yes
Blueberry, half-high	4 to 7	Acid	Light	Late summer/fall	Yes
Cranberry	2 to 6	Acid, moist	Light	Fall	No
Currant, black	3 to 6	High fertility	Moderate	Summer	Yes
Currant, red & white	3 to 6	High fertility	Moderate	Summer	No
Elderberry, American	2 to 9	Deep, drained	Light	Late summer/fall	Yes
Gooseberry, European	3 to 7	High fertility	Moderate	Summer	No
Gooseberry, American	3 to 7	High fertility	Moderate	Summer	No
Jostaberry	3 to 7	High fertility	Moderate	Summer	No
Serviceberry	3 to 8	Well-drained	Light	Summer	No

Pruning to a Leg

Currants, gooseberries, and jostaberries all thrive when they are grown as a "leg"—a single stem that you train to act as a trunk. This method is easy to do, but takes more time each year than growing plants as simple bushes.

1. After developing the central stem and allowing four good branches to grow, head them back.

2. Each year, when the plant is dormant, head back the branches, and remove any branches from the trunk.

3. To promote good growth, head back the tips of laterals on which fruit forms each year.

4. Remove 4-year-old branches each year, and thin twiggy and poorly positioned growth.

Pruning Currants, Gooseberries, and Jostaberries

These plants of the *Ribes* species can be grown as bushes or "stools" on a single stem, or "leg," and except for black currants, as espaliered cordons with vertical fruiting branches.

Growing these plants as bushes is by far the easiest alternative. If you are a beginning fruit grower, this is probably the best choice. To grow the plants as stools, cut back all of the stems to two to four buds above the soil level just after planting. During the first year, the plant will develop branches from each of these buds, as well as new branches that originate underground.

When the plant is dormant, thin out any weak growth, and then thin the strong, upright growth so that only six or seven stems remain. Thin stems by cutting just at or slightly below soil level. In subsequent years you will repeat this dormant pruning, but in the third year you must thin out all of the three-year-old stems.

Pruning on a Leg

To prune plants to grow on a leg, rub off all the buds on the bottom 6 inches or so of the central stem. Prune the plant during the dormant season almost as you would if it were a tree you were growing in a central-leader design. First develop four or five branches as a permanent framework, and then head back the lateral branches that form on the framework when you prune each year. Thin out weak branches and twiggy growth in the center of the plant each year, and remove branches that are more than 3 years old.

Lowbush, Rabbiteye, and Highbush Blueberries; Red and White Currants

Fruiting Shrubs, Plant by Plant

BLUEBERRY

While best known in their raw form, blueberries can be dried, frozen, and made into sauce.

LOWBUSH, RABBITEYE, AND HIGHBUSH BLUEBERRIES

Zones: Lowbush, 3 to 8; rabbiteye, 6 to 9; highbush 4 to 8; and half-high 4 to 7.

Size: Lowbush, 1 to 3 feet tall, spread 5 to 8 feet wide; rabbiteye, 15 to 18 feet tall, 5 to 6 feet wide; highbush, 5 to 6 feet tall and wide; half-high, 2 to 4 feet high and wide.

PLANTING

Exposure: Half-high and highbush blueberries need full sun. Rabbiteyes can tolerate filtered shade in the afternoon in hot locations. Lowbush cultivars prefer afternoon filtered light or moderate shade.

Soil: Consistently moist with high organic matter and extremely good drainage. The pH should range between 4.0 to 5.0, although rabbiteyes will tolerate 5.5.

Spacing: Allow 8 feet between rows of lowbush, highbush, and half-highs and 5 to 6 feet between plants. The larger rabbiteye plants require a spacing of 8 feet in all directions.

Pollinators: For pollination, plant at least three different cultivars.

CULTURAL CARE

Watering: Lowbush, half-highs, and highbush plants require 1 to 1½ inches of water a week. Rabbiteyes can tolerate 1 inch per week.

Mulching: Apply a high-acid mulch, such as pine needles or shredded hardwood tree leaves.

PRUNING

After Planting: Prune as little as possible, but remove flower buds.

Routine Pruning: In dormant season, thin the center of plant.

Pruning Mature Plants: Remove branches older than 3 years on half-high, highbush, and rabbiteye plants. Cut half the plants of a lowbush planting to the ground each year, alternating sections of the planting. Do not prune the tips of rabbiteye stems; fruits form here.

HARVESTING

Fruiting: Fruits form on 1- to 4-year-old branches or spurs.

Years to Bearing: 4 to 5.

Harvest Season: Late summer to fall.

Yields: 3 to 8 quarts for all most types; 2 to 4 quarts for lowbush.

RED CURRANT

Red currants have a sharp flavor; white currants are a touch sweeter. Both can be made into jam, jelly, and juice. One warning: white pine blister rust depends on both black currants and pines for its life cycle and some states outlaw all Ribes species—red and white included. Check with your state department of agriculture.

RED AND WHITE CURRANTS

Zones: 3 to 6.

Size: 3 to 5 feet tall, 3 to 5 feet wide.

PLANTING

Exposure: Full sun in Zones 3 to 5, but filtered afternoon light in Zone 6.

Soil: Moist, well-drained, slightly acid soil with high organic matter. Each year, fertilize with 3 pounds of soybean meal.

Spacing: 4 to 6 feet between plants, 6 to 8 feet between rows.

Pollinators: None required.

CULTURAL CARE

Watering: A minimum of 1 inch a week or 2 gallons per square foot of root area.

Mulching: 6 inches of straw or 3 to 4 inches of a denser material.

PRUNING

After Planting: Head back to 2 to 4 buds.

Mature Plants: Maintain design.

HARVESTING

Fruiting: Fruit forms on wood that is between 1 and 3 years old.

Years to Bearing: 2 to 3.

Harvest Season: Midsummer. Taste the bottom berry on a cluster to determine stage of ripeness.

Yields: 3 to 5 quarts per bush.

Growing Fruit, Berries, and Nuts

3

American Elderberry; Saskatoon Berry, Serviceberry, Juneberry

ELDERBERRY

While these can be eaten when fully ripe, they contain toxins that are fully eliminated with cooking. They need some sweetening when cooked in pies. Wine and jam are also popular elderberry products.

AMERICAN ELDERBERRY

Zones: 2 to 9.

Size: 6 to 12 feet tall, 5 to 6 feet wide.

PLANTING

Exposure: Full sun in Zones 2 to 5, filtered afternoon shade in Zones 6 to 9.

Soil: Deep, fertile, well-drained soils with high levels of organic matter.

Spacing: 5 to 6 feet between plants.

Pollinators: Plant two cultivars.

CULTURAL CARE

Watering: 1 inch per square foot of root area each week.

Mulching: 6 inches of straw or 3 to 4 inches of a denser material.

PRUNING

After Planting: Prune as little as possible.

Routine Pruning: Keep center of plant open. Thin out weak or diseased growth.

Pruning Mature Plants: Thin out branches older than three years, and remove suckers.

HARVESTING

Fruiting: Fruits form on tips of 1-year-old wood and on 2-year-old branches.

Years to Bearing: 2 to 4.

Harvest Season: Late summer to midfall.

Yields: 12 to 15 pounds.

SASKATOON BERRY

Many people find the serviceberry comparable in flavor to the blueberries. This grape-size fruit is juicy and sweet whether eaten fresh or cooked in a dessert.

SASKATOON BERRY, SERVICEBERRY, JUNEBERRY

Zones: 3 to 8.

Size: 6 to 40 feet tall, 4 to 20 feet wide.

PLANTING

Preferred Exposure: Full sun but will tolerate partial shade, particularly in the southern reaches of their range.

Soil: Well-drained of moderate fertility and organic matter content.

Spacing: As far apart as the mature plant will be tall.

Pollinators: None required.

CULTURAL CARE

Watering: Each week, 1 inch or 2 gallons per square foot of root area.

Mulching: Mulch with 6 inches of a light mulch, such as straw, or 4 inches of a denser material.

PRUNING

After Planting: Prune as little as possible.

Routine Pruning: Prune bushy types by thinning out new growth during the dormant period to keep the center of the plant open. Remove weak growth.

Pruning Mature Plants: Prune to keep plants within bounds by thinning out suckers that travel too far from the mother plant. Head back drooping branches, and cut out growth more than 3 years old.

HARVESTING

Fruiting: Fruits on 1-year-old wood.

Years to Bearing: 2 to 4.

Harvest Season: Mid to late summer.

Yields: 4 to 6 quarts per plant.

Choosing Grapes

Grapes can be white, red, or purple and meant for eating out of hand, juicing, turning into wine, or drying for raisins. The huge, flavorful muscadine grapes are common in the South, while California hosts most of the European and hybrid cultivars. American grapes (*Vitis labrusca*) are grown in moderate climates from coast to coast. Grapes are the most widely grown fruit in the world, but in most areas only one or two types are raised because of climate and soil constraints. Choosing species and cultivars appropriate to your climate is important to your success with any crop, but this is especially true with grapes. Climate can determine disease resistance as well a good reason to buy local.

American grapes can tolerate conditions as far north as protected spots in Zones 4 and 5. The best-known cultivars that can withstand Zone 4 conditions are 'Canadice', 'Concord', and 'Reliance Seedless'. All three are resistant to some common diseases, and both 'Canadice' and 'Reliance Seedless' are red. The large white 'Niagara' and blue 'Mars' grow well in Zone 5 but do not fare well in Zone 4.

Muscadine grapes are also native to the United States but are a different species than the grapes we call 'American'. Muscadines grow well from Zones 7 to 10, primarily in the Southeast. Their resistance to pests and diseases makes them able to thrive in high humidity. Unlike other types of grapes, some of the best muscadines require a pollinator to set fruit. The vines commonly grow more vigorously too, so growers usually train them on a Munson trellis, as described on page 110.

The most cold-resistant cultivar is 'Carlos'. It will tolerate protected spots in Zone 6b, although it grows better in Zone 7 southward. 'Scuppernong' and 'Hunt' are well-known cultivars throughout the South.

European grapes grow well in Zones 7 to 9, primarily in the lower-humidity conditions in California and the Southwest. Many of these grapes are used to make wine, but some, such as 'Thompson Seedless', are good table grapes.

In the late 1880s, the grape phylloxera aphid attacked French vineyards, but not American grapes. Breeders crossed European and American grapes. The resulting hybrids are grown as far north as Zone 5. 'Himrod' and 'Lakemont Seedless' are among the most popular because of their cold tolerance and because they have some of the disease resistance of their American parents.

With a little training, grapes will fully cover a garden pergola.

Grapes can do well in most any region as long as you select the species and cultivar carefully.

107

Pruning and Training Grape Vines

Grape vines are beautiful plants for covering arbors and roofs. Grown this way, the plants can give high yields of healthy fruit if you keep up with pruning and training. And beware of high humidity levels. Some disease-resistant cultivars tolerate such conditions well, but others fall prey to fungi such as anthracnose or black rot.

Grapes are labor intensive, requiring vigorous annual pruning and training. The easiest and most common way to trellis European, hybrid, and American grapes is with the four-armed Kniffen system.

Set up the trellis before you plant. Use exterior-grade 4×4 or 6×6 posts, burying them below the frost line. Extend them 6 feet above the soil surface, and space them 6 to 8 feet apart. Stretch 12-gauge wires from post to post at about 2 to 3 feet from the soil surface and at the top of the posts. Anchor the wires to strong stakes at each end of the trellis.

Plant each grape directly under the wire in the middle of the space between two posts. Prune back the vine to two buds near the bottom wire, and pinch off all the other buds. As the shoots develop from the two buds over the season, tie them to the wire without letting them twine around it.

The following late winter or early spring, while the plant is still dormant, select the shoot that seems the strongest and most vigorous to become the trunk of the vine. Untie it from the bottom wire, and lead it upward to the top wire. Tie it to that support. Prune off the other shoot and any lateral branches that have formed, leaving two buds near both the top and bottom wires.

Allow four buds, two at the bottom wire and two at the top, to develop during the summer, but pinch off all those that grow from the top and center portion of the trunk. Allow a renewal bud to grow on each shoot near the wire.

During the dormant period, head back the shoots, or canes, to about 10 buds. During the season, grapes will grow near the bases of shoots that grow from these buds. Allow the renewal buds to develop, too, because this is where fruiting shoots will grow the following year.

In subsequent years during the dormant period, cut back the canes where fruit formed the year before; head back the current year's fruiting canes to ten buds; and allow renewal buds to remain on the plant.

Muscadine grapes are generally grown with a Munson system. This style allows greater air circulation around the foliage, making it particularly appropriate for grapes grown in the humid Southeast. Set up a T-bar Munson trellis, as illustrated on page 110, with posts set 6 to 8 feet apart and T's 3 feet long. Secure the 12-gauge wire 4 feet above the soil surface and on the top surfaces at the ends of the T-bars. Plant the grapes under the center wire, between two posts, and cut the plant back to two buds. Allow these to grow during the season. While the plant is dormant, choose the shoot that will become the trunk

Grapes do well on an overhead arbor, an efficient use of space if you need shade over a deck or patio. However, be sure the harvest the grapes before they drop.

of the grape plant; lead it to the wire; and tie it in place.

Allow two shoots to develop from the trunk, and tie them to the central wire as they grow over the summer. During the dormant period, head back the canes. Allow two renewal buds to remain on the plant for the following year, and prune back any other growth.

During the season, shoots will grow from the arms. Drape them over the top wires. Branches that grow from these shoots will form fruit the following year. Each year, allow two new shoots to form for the following year's fruit.

If you are growing a grape over an arbor, start it off as if you were growing it as a four-armed Kniffen system. However, rather than allowing fruiting canes to develop low on the trunk, pinch them off in favor of those growing near the top of the vine. Each year, allow renewal buds to form near the supporting trelliswork of the arbor. Prune off all the others.

Training Grapes

European, American, and hybrid grapes are usually pruned to a four-armed Kniffen system. However, if space is tight, you can also grow them as a two-armed plant, or cordon as shown. The basic pruning is the same whether you leave four arms or two.

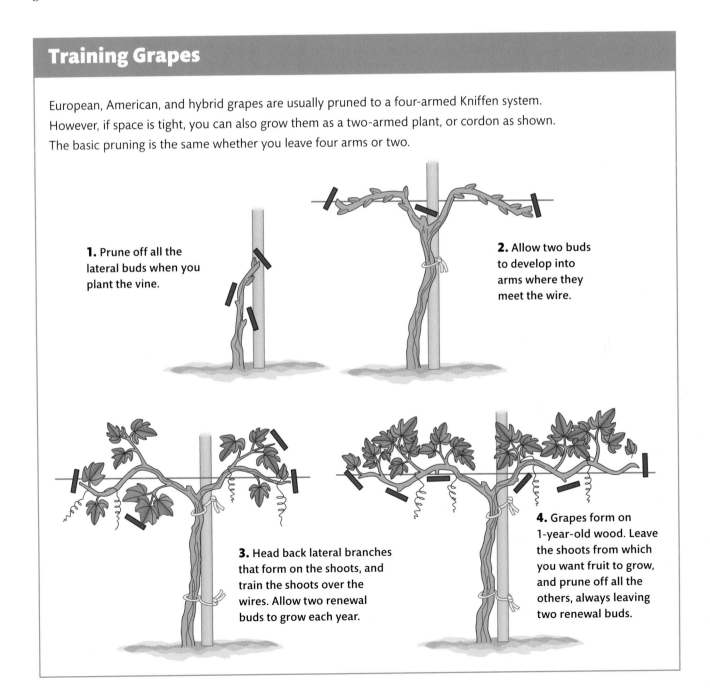

1. Prune off all the lateral buds when you plant the vine.

2. Allow two buds to develop into arms where they meet the wire.

3. Head back lateral branches that form on the shoots, and train the shoots over the wires. Allow two renewal buds to grow each year.

4. Grapes form on 1-year-old wood. Leave the shoots from which you want fruit to grow, and prune off all the others, always leaving two renewal buds.

Training Muscadine Grapes

Muscadine grapes are pruned and trained to a Munson system, primarily because the vines are heavy. This system allows greater air circulation, reducing the plants' vulnerability to the fungal diseases that are so prevalent in the humid regions where they grow. Prune the vine to two buds near the bottom wire after planting, and pinch off all other growth. Tie the shoots that develop to the bottom wire.

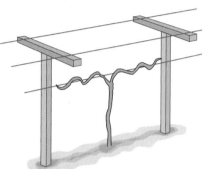 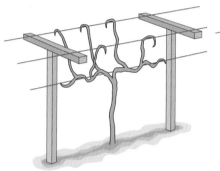 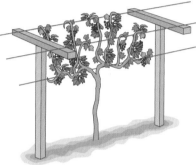

1. Stretch 12-gauge wire about 4 ft. above the soil and over the tops of the Ts. Tie the first-year's growth to the bottom wire.

2. Drape the shoots that develop the following year over the top wires, spacing them well. Fruit will form on the laterals in the following year.

3. During the summer each year, allow two new shoots to form, and tie them to the support. Laterals growing from them will provide fruit the next year.

Caring for Your Vines

Caring for your grape vines is part of the pleasure of growing them. Check them daily to be certain that they've had enough water, that growth is staying within bounds, and that all the growth looks healthy.

Grapes require high levels of balanced nutrition. If you use compost tea foliar sprays every month during the active growing season, you can add some nutrients and prevent many of the diseases that attack fruit vines. Similarly, liquid seaweed foliar sprays can supply trace elements and help to protect the plants against light frosts.

Foliar sprays will help keep the vines healthy, but even so, it is best to check the nutrient status by sending leaves to a lab for tissue analysis every 3 years. Let the lab know that you are working with organic fertilizers so that they will give you appropriate recommendations.

All through the season, monitor your vines closely for pests and diseases. If you catch these problems in the beginning, your control job will be easier and more pleasant. Check the fruiting vines directory, pages 111 and 112, for care of grape vines.

One of the best pest- and disease-control measures is good sanitation. As leaves drop in the autumn, rake them up and compost or destroy them. Protect the vines with screen mesh and white latex paint, and add new mulch over the root area once the top inch of the ground has frozen.

If you bury your vines for winter protection, dig the trench long before the ground is frozen. Wait until all the leaves have dropped, and rake them up from the area. Untie the vine carefully, and gently lower it into the trench. You may have to spade around one side of the root-ball to give the vine a way of bending over. If so, pile a foot of soil over the upended roots, and then cover them with mulch.

Grape Vines, Plant By Plant

AMERICAN GRAPES

Also known as the Fox grape, this species includes the famous Concord grape. Some of the new American cultivars have thinner skins than traditional Concords. When you pick for jams or jellies, gather the fruit before it is fully ripe. Under-ripe fruit makes a more flavorful product and contains high levels of natural pectin.

AMERICAN GRAPE

Zones: 5 to 9.

Size: 12 to 15 feet long, 3 to 5 feet wide.

PLANTING

Exposure: Full sun, preferably on a south-facing slope with good air circulation and drainage.

Soil: Deep, well-drained soil with high fertility and organic matter content. Add ½ to 2 pounds of compost per row foot in early spring each year.

Spacing: Allow 8 feet between plants, 9 feet between rows.

Pollinators: None required.

CULTURAL CARE

Watering: Plants require 4 gallons per square foot of root area every week from early spring until the fruits begin to color. After that, hold back on watering until the end of the season unless natural rainfall is less than ½ inch per week. In that case, water deeply every 2 weeks.

Mulching: Rake off old mulch, along with fallen leaves, at the end of the season, and mulch for winter protection. Remove winter mulch as soon as all threat of frost has passed.

PRUNING

See page 108–109.

HARVESTING

Fruiting: Fruits form on 1-year-old wood.

Years to Bearing: 2 to 3.

Harvest Season: Late summer to early fall, depending on cultivar.

Yields: 10 to 15 pounds per plant.

EUROPEAN GRAPES

Brought to Mexico in the 1500s by the Spaniards, the European grape made its way to California. The rest is wine-making history. European grapes have just the right amount of natural fruit acids and sugars to make red and white wine and most everything in between.

EUROPEAN AND HYBRID GRAPES

Zones: 7 to 9.

Size: 12 to 20 feet long, 4 to 5 feet wide.

PLANTING

Exposure: Full sun in an open site.

Soil: Deep, rich soil with high levels of organic matter. Apply 2 pounds of compost per row foot in spring.

Spacing: Allow 6 to 8 feet between plants and 10 feet between rows.

Pollinators: None required.

CULTURAL CARE

Watering: Supplement natural rainfall if necessary to supply 2 inches, or 4 gallons per square foot of root area, of water each week from spring until the fruits start to color.

Mulching: Do not mulch for the first summer. For winter protection, pile 8 to 12 inches of straw or 4 to 6 inches of a denser material over the roots once the top inch of the ground has frozen. In spring, remove the mulch, and do not add new material until midseason. Remove and compost old mulches when you rake up the leaves in autumn, and mulch again for winter protection.

PRUNING

See pages 108–109.

HARVESTING

Fruiting: Fruits form on 1-year-old wood.

Years to Bearing: 2 to 3.

Harvest Season: Late summer to fall.

Yields: 10 to 30 pounds per plant.

111

Growing Fruit, Berries, and Nuts

MUSCADINE GRAPES

This American native flourishes in the warm and humid southeastern states where American and European do poorly. The Muscadine is best for jellies, jams, juices, and dessert wine.

MUSCADINE GRAPE

Zones: 7 to 10.

Size: 12 to 20 feet long, 3 to 5 feet wide.

PLANTING

Exposure: Full sun.

Soil: Deep, well-drained, fertile with high levels of organic matter. Apply 1 to 2 pounds of compost per row foot every spring.

Spacing: 15 to 20 feet between plants and 15 feet between rows.

Pollinators: Check with the nursery for appropriate pollinators, and grow the plants within 50 feet of each other because they are wind pollinated.

CULTURAL CARE

Watering: Supply 2 inches or 4 gallons per square foot of root area every week from spring until the fruit begins to color. Do not water after this unless there is a severe drought. Do not wet leaves when watering.

Mulching: Do not mulch the first season until late fall. Remove mulch in spring, and allow the soil to be bare until midseason. After that, mulch with 6 inches of loose material or 4 inches of a dense substance. Remove mulch in fall, and apply new material for winter protection.

PRUNING

See pages 108–110.

HARVESTING

Fruiting: Fruits form on 1-year-old wood.

Years to Bearing: 2 to 3.

Harvest Season: Late summer and fall.

Yields: 8 to 16 quarts per plant. Muscadine grapes form looser clusters than American or European cultivars.

Taste the bottom grape on a cluster before cutting it from the vine because grapes do not ripen once they are picked.

A purpose-built trellis is ideal for growing grapes because it gives you ready access to grapes at harvest time and makes it easy to prune the vines properly. However, grapes can also flourish on a fence, arbor, or lattice panel.

Choosing Bramble Fruit

Bramble fruits are extraordinarily easy to grow in most locations and yield large amounts of fruit for the small amount of space they take up. Despite their susceptibility to various pests and diseases, you can count on 8 to 15 good years from a single planting if you take minimal care of them.

The table on page 115 lists only a few of the more than 250 known bramble species. Despite their wide adaptation, they are picky about their environments. Check with local nurseries, gardeners' associations, and neighbors to learn which cultivars are best for your area.

Summer or Fall Bearing

Bramble bushes grow a new set of canes, or stems, from the crown each year. Those produced in the current year are called *primocanes*. When they've overwintered and resumed growth in the spring they are called *floricanes*.

Lateral branches growing from the floricanes produce the flowers and fruits on most brambles. However, everbearing or fall-bearing types set fruit on the tips of the current year's primocanes as well as on laterals growing from the floricanes. 'Heritage', 'Fall Red', and 'Fall Gold' are all well-known fall-bearing raspberries. Include both summer- and fall-bearing plants on your farmstead.

The densely clustered berries of bramble fruits provide a rich annual harvest. They are so tasty when ripe that many pickers stipulate "one for the mouth, one for the bucket."

Raspberry or Blackberry?

Aside from flavor, an important difference between raspberries and blackberries is the way the fruit separates from the bush. When you pick a raspberry, the central core of the fruit stays on the bush. In contrast, the central core comes along with a harvested blackberry. In fact, a ripe blackberry almost falls into your palm as you grasp it. Another difference is hardiness. Raspberries are more robust than blackberries, growing well in Zones 3, 4, and the coldest areas of 5a, spots where blackberries suffer during cold winters. All raspberries are selffruitful, but some blackberries require pollinators.

The bramble fruit that survives being eaten fresh can be made into jam and jelly or frozen to be enjoyed later.

3 Growing Fruit, Berries, and Nuts

Siting Your Brambles

If possible, site the plants where late frosts are unlikely and winter winds are moderated by trees or buildings. Gardeners living in Zones 5a to 3 who have no choice but to plant the brambles in cold, exposed frost pockets may need to wrap the canes in burlap.

Bramble fruits remain healthy and productive for 10 to 15 years, but they eventually lose vigor because of the cumulative effect of the various viral diseases that prey on them. When you replant the fruit, move them to an entirely different part of the yard so that the new bushes can avoid the diseases.

Planting, Pruning, and Training Brambles

Bramble fruits grow best on a trellis. The three most common types of trellises for brambles include the T-bar trellis, the double-post-and-wire trellis, and the cable-and-post-fence support. Each of these supports can be used for any of the erect or semi-erect cultivars, but the cable-and-post-fence support is the most practical for the long canes of trailing blackberries.

Set up the trellis you plan to use before planting the berries. Plant the berry bushes slightly deeper than they were growing at the nursery. Space them as recom-

mended in the bramble fruit directory on page 116. If the tip of a cane looks damaged, head it back until you reach healthy tissue. Head back undamaged canes to a length of 6 inches to stimulate new growth.

During the first year, summer-bearing cultivars will produce a healthy crop of primocanes—the stems that will become fruiting floricanes the following year.

Your job during the first year will be to move the canes into position between or against the wires and tie them in place. During the dormant season, thin the canes if they are growing more closely together than 6 inches, and head back the remaining canes. If you are growing a trailing blackberry, wrap the long canes around the top wire and head them back only a few inches. Head back the shorter, erect cultivars so that they are only a foot above the top wire.

During the summer, tie the new primocanes to the bottom wire of your support until after the floricanes are harvested. Or divide the planting so that all the floricanes are trained to one side and all the primocanes to the other. Separating the canes in this way saves time and trouble in the long run, especially when you are picking and pruning old floricanes.

Once the floricanes have fruited, they will die. Thin them, making the cuts just above the soil surface. Remove the old canes from the area, and do not compost them. Retie the primocanes.

Propagating with Suckers

Bramble fruits propagate themselves by forming suckers from their far-ranging roots. If they do not have the viral diseases that aphids often transmit, you can transplant these to add to your planting.

In early spring, as soon as you can identify the suckers, dig them up, left. Take as large a root-ball as possible and transplant as usual, right.

Brambleberries

COMMON NAME	NOTABLE CULTIVARS	ZONE	NOTES
Asian Raspberry		9 to 10	Best for Florida and other Zone 9 areas
Crimson Bramble		1 to 4	Dark red fruit on truly cold-hardy plants
American Dewberry		5 to 9	Purple fruit on nearly thornless canes
European Red Raspberry		3 to 8	Includes both red and yellow types
American Red Raspberry		4 to 7	Includes both red and yellow types
	'Latham'	3 to 8	July fruiting
	'Dorman red'	6 to 9	Best for southern conditions
	'Canby'	5 to 8	Nearly thornless, aphid immune, July fruit
Black Cap Raspberry		3 to 8	American native, some are purple
	'Royalty'	4 to 8	A purple hybrid of American & Black Cap
	'Black Hawk'	4 to 8	Somewhat virus resistant
Wineberry		5 to 8	Fine flavored, large fruit on erect canes
Blackberry, Erect		5 to 10	Many types are available
	'Darrow'	5 to 10	Best in Northeast
	'Dirksen'	5 to 10	Rust resistant
	'Cheyenne'	6 to 10	Best for Gulf Coast and South
Blackberry, Trailing		5 to 10	Must always be trellised
	'Cascade'	5 to 8	Best for Pacific Coast
Thornless Blackberry		5 to 10	Grows best in southern zones
	'Perron Black'	5 to 8	Adaptable but does best from Zones 6–8
Loganberry		5 to 9	A cross between raspberry and blackberry
Boysenberry		6 to 8	Derived from the hybrid loganberry

Watering and Routine Care

After you plant, soak the area thoroughly. Plants should receive at least 1 inch a week, or 2 gallons per square foot of root area. Do not irrigate while the fruits are developing unless there are drought conditions. If you do not have a drip system, hand water rather than use a sprinkler. This allows you to direct the water to the soil rather than the plants. Spray plants with compost tea on a regular schedule to keep fungal problems at bay. Check for signs of pests and diseases on a regular basis.

3 Growing Fruit, Berries, and Nuts

Blackberry, Raspberry, Boysenberry, Loganberry, Dewberry

Bramble Fruits, Plant by Plant

BLACKBERRY

Jams, pies, and dessert toppings are popular uses for these berries.

BLACKBERRY, RASPBERRY, BOYSENBERRY, LOGANBERRY, DEWBERRY

Zones: Blackberry, Dewberry, and Loganberry, 5 to 10; Boysenberry, 6 to 8; Raspberry, 3 to 9.

Size: Erect cultivars, 5 to 10 feet long, 3 to 5 feet wide; Trailing cultivars: 15 feet long, 4 to 5 feet wide.

PLANTING

Exposure: Full sun. Protect from winds in Zones 6 and cooler.

Soil: Deep, well-drained soil of average fertility and high organic matter. Apply 1 to 1½ pounds of compost per row foot in early spring each year.

Spacing: Erect cultivars, 3 to 5 feet between plants; trailing cultivars, 8 to 10 feet between plants.

Pollinators: Occasionally required. Check with your supplier.

CULTURAL CARE

Watering: Plants require 1 inch of water a week from spring to the time the fruit is coloring. Unless there is a severe drought, do not supplement natural rainfall from the time the fruit begins to color until it is harvested.

Mulching: Keep plants mulched to prevent weeds from growing in the area. Use newspaper or cardboard covered with straw or shredded hardwood bark. Remove mulch each fall, and replace with clean material to help keep diseases in check.

PRUNING

After Planting: Head back canes to 6 inches.

Routine Pruning: See page 114.

Fruiting: Most brambles fruit on 1-year-old canes; some raspberry cultivars fruit in fall.

Years to Bearing: 1.

Harvest Season: July to fall, depending on species. Taste-test for ripeness and harvest every other day during the season.

Yields: 1 to 8 quarts per plant, depending on species.

Boysenberries are a stabilized hybrid with blackberry, loganberry, and raspberry parents. They resemble blackberries and loganberries more than they do raspberries. They differ from blackberries in that they require a spacing of 8 feet between plants, they never require a pollinator, and their larger vines yield up to 8 quarts per plant.

Dewberries are one of the most distinctively flavored of the brambles—sweet with tart undertones. 'Carolina' is a cultivar that resists fungal leaf diseases.

Loganberries are tart and an acquired taste for many people. Loganberries resemble raspberries more than blackberries when it comes to the pests and diseases they attract.

Training brambles onto fences or trellises keeps them accessible and off the ground.

Strawberries

Almost everyone is familiar with the traditional June-bearing strawberries. True to their name, they fruit just once a year, generally in June.

Ever-bearing (sometimes called day-neutral strawberries) have become more popular during the last few years. Rather than bearing all their fruit at once, they produce small amounts all through the season with peaks in June and August. By combining the June-bearing and ever-bearing, you can rely on the June-bearers for bumper crops to preserve and freeze and a patch of day-neutrals for a steady supply through the season.

Alpine strawberries are also ever-bearing, but their fruits are the size of a baby's thumbnail. Their fruit tastes more like wild strawberries. Alpine strawberries can be grown from seeds rather than plants and thrive in hanging baskets.

Strawberries are one of the easiest crops you can grow. However, to get the best possible fruit, you'll need to plan to move the crop every four or five years. Grow a nongrass cover crop in the area where you plan to grow your next crop of strawberries to avoid building high populations of cutworms or beetle grubs.

Choosing Cultivars. Read the cultivar descriptions, right, carefully; choose those with fine flavor rather than large size or good shipping qualities—those are the characteristics that make for the pulpy stuff you find at the supermarket. In northern areas, you may also want to choose mid- and late-season types so you don't have to worry about protecting the plants from late spring frosts.

Versatile strawberries can be raised in patches or pots. A combination June-bearing and day-neutral plants will extend your harvest through the summer.

Favorite Strawberry Cultivars

	ZONES	
ALPINE CULTIVARS		
'Baron Solemacher'	4 to 8	Grows well in pots anywhere
'Ruegen Improved'	5 to 8	The alpine most widely grown commercially
'Pineapple Crush'	5 to 8	A yellow fruit with good flavor
DAY-NEUTRAL CULTIVARS		
'Aptos'	6 to 9	Commonly grown in California and Arizona
'Fort Laramie'	3 to 8	A good choice for the North
'Ozark Beauty'	5 to 8	This plant is fairly disease-resistant and yields well
'Ogallala'	5 to 8	Tolerates dry conditions better than most
'Tristar'	5 to 8	Known for good flavor
'Tribute'	4 to 8	Resists most diseases and has good flavor
JUNE-BEARING CULTIVARS		
'Earliglow'	4 to 8	Disease resistant and matures early
'Cardinal'	4 to 8	Matures early and is common in the mid-Atlantic states
'Dunlap'	4 to 8	Tolerates a range of soils, including clays
'Kent'	4 to 8	Hardy to Zone 3 if protected; yields early
'Honeoye'	4 to 8	Widely grown commercially because it is disease resistant and has huge fruit; some consider it a bit bland
'Red Chief'	4 to 8	Hardy to Zone 3 if protected; resists most diseases, and matures midseason
'Sparkle'	4 to 8	A fine-flavored, late-season berry

Nuts

Like fruit trees, nut-bearing trees are an investment in the future—in some cases a long-term investment. A hazelnut tree offers the quickest return, producing nuts in as little as 2 years. At the other extreme, a walnut tree will take 10 to 12 years before you have a useful harvest of nuts. Some require a lot of space. However, a nut tree can be an attractive asset, bearing blossoms in the spring and beautiful foliage during the growing season—a great way to make productive use of a front yard.

Most nut-bearing trees are low maintenance, requiring only pruning at the end of the season. Here are your prime options for growing edible nuts.

Hazelnut

Also known as the filbert, the hazelnut is compact, growing 6 to 20 feet tall, and is bushlike enough to make a hedgerow. It grows in Zones 3–9, and can live for 40 years.

Choose two varieties to ensure cross-pollination. Plant in fertile, well-drained soil. It takes 2 to 4 years for plants to start bearing.

The nuts grow inside a tough husk. At harvest time, husks will begin to fall, your cue to start your harvest. Gently shake the branches to release the husks. Spread them in the sun to dry for a week or two. The husks can then be removed from the nut by hand. Still in the shell, load the nuts into a mesh bag and store at 34–40°F.

Almond

Almond trees grow in Zones 7–9. Some hardy varieties can handle Zone 6 as well. Dwarf almond trees are only 8 feet tall; full-size trees are no more than 15 feet tall.

You'll need to wait about 4 years for your almond trees to bear, and 12 years for them to come into full production. Almonds are not self-pollinators.

In August, September, or October the almond hulls will begin to split, signaling harvest time. Choose a period when dry, sunny weather is predicted. Rake the area beneath each tree. Knock the nuts off the tree by thrashing the limbs or shaking the tree. (Don't shake young trees.) Allow the hulled almonds to dry on the ground for two days or more, and then remove the split hulls. Store the almonds with or without their shells in the freezer.

Chestnut

The Chinese chestnut accounts for most of the chestnuts raised in the U.S. The tree grows 50 feet tall and can produce in 6 to 7 years.

The Chinese chestnut is suitable for Zones 4–8, but some hybrids can handle the frigid extremes of Zone 3. A chestnut tree needs well-drained soil and must be protected from grass and weeds that compete for moisture, nutrients, and root growing space. Establishing a ground cover of bluegrass or clover is beneficial, as is mulch for weed control.

In September the spiny husks that hold the chestnut begin to open. They'll fall to the ground of their own accord. Chestnuts need to be kept in a cool place with just enough moisture to keep them from drying out.

Pecan

Pecan trees are big—they can top out at 130 feet—and they need 35 to 50 feet between plants. Pecan trees are prone to insect and disease problems, which commercial growers spray for as many as 10 times a year—a difficult task for the small-scale farmer that may involve undesirable pesticides and fungicides. Organic alternatives include biological controls using beneficial insects, and low-toxicity pesticide and fungicide spraying. If you plant new trees, you'll have to wait 5 to 10 years for them to produce nuts.

Pecan trees flourish in the long, warm growing season of Zones 6–9. They require plenty of water and need soil that drains well. They may require additional irrigation.

Harvest runs from October to December. A padded stick is enough to jar limbs and cause the nuts, still in their hulls, to fall to the ground. After removing the hull, pecans are best stored in their shell or shelled and frozen.

Hickory

Related to the pecan, hickory nuts are tasty but have very hard shells that require a good whack with a hammer to split, a dicey business because hickory nuts are smaller than a ping-pong ball. Once cracked, the meat has to be removed with a nut pick—one of the reasons why their popularity is limited.

Hickory trees are great shade trees, slowly growing to 80 feet in height. They grow in Zones 5–9 and require 10 years before they start to produce nuts. The hull-covered nuts fall on their own accord. Remove the hull, and drop the nuts in a bucket of water. Edible nuts will sink; discard floaters. Dry the nuts, and store them in a cool, dry place.

Walnut

The familiar store-bought cracking walnut is the English walnut. Its darker, more pungent relative, the black walnut, is sold in chopped bits because it is difficult to remove the meat from its tough shell. Both come from trees that may top out at over 80 feet and spread 60 feet. They are hardy, growing in Zones 4–6. It takes walnut trees 10 to 12 years to start producing.

Of the two types of walnut trees, the black walnut is the most problematic. On the plus side, its wood is valuable as veneer and can be harvested in 25 years. On the negative side, its nut husks are tough and loaded with tannins that stain everything they come in contact with. It is difficult to get at the meat inside. Also, the roots produce a substance called juglone that is toxic to a number of plants.

The English walnut is easier to handle. After rinsing to remove tannins, the shells are air-dried for a few weeks and then cracked.

Both types provide a good harvest about two out of every five years. The husked nuts are harvested by hooking a limb and shaking it until they fall—a hard-hat job. Store in a cool, dry place.

119

4

Raising Chickens

THE CHICKEN IS NO ONE'S NATIONAL BIRD

(although Rhode Island proudly claims it as state bird), seldom figures on family crests, and is the butt of many jokes. Hardly fair treatment of an animal that has been a farmyard mainstay for thousands of years and may just be the perfect backyard animal. Not only do chickens produce eggs that are easy to store and famously nutritious, they devour insect pests, relish kitchen and garden waste, and deliver highly fertile manure. And when their productive life is over, they can even fill the stew pot. All this while occupying less space than a hot tub. That's why so many people are making the chicken choice.

"Other people keep cats and dogs," says Cleveland flock owner Ben Shapiro. "But cats don't mow your lawn and make eggs from your table scraps. And chickens are fun to watch. I love them. For me they're more entertaining than television."

Changing Codes

At the turn of millennium you'd have been hard put to find a municipality that allowed chickens at all. Today, cities such as New York, Los Angeles, Chicago, Denver, Miami, Minneapolis, and Seattle—to name a very few—allow backyard chickens to one degree or another. Restrictions vary somewhat, but regulations are consistent in allowing only egg-type birds, limiting the number of hens, prohibiting roosters, and requiring that layers be penned rather than have the run of the property. (See the table below for a sampling of regulations.)

Because they are quiet and easy to care for and don't require a lot of space, chickens make good city dwellers. Stephanie and T. J. Johnson of Olympia, Washington, consider them a key ingredient in their self-sufficient lifestyle.

Urbane Chickens

If a small flock of hens interests you, the first step is finding out if they are legal in your area and, if so, whether or not you can house them far enough away from neighbors. Here's a sampling of chicken regulations.

CITY	BIRDS	SETBACK	SPECIAL REQUIREMENTS
Baltimore, MD	4 hens	25 ft. from neighbor's house	Register with animal control
Charlotte, NC	Based on lot	25 ft. from property line	$40 fee, minimum area, 4 sq. ft. per bird
Houston, TX	7 hens	100 ft. from neighbor's house	Minimum lot 65 ft. x 125 ft.
Missoula, MT	6 hens	20 ft. from neighbor's house	$15 fee, rat-proof feed containers
Seattle, WA	3 hens	50 ft. from property line	$40 fee, 1 additional hen per 1,000 sq. ft. of lot
So. Portland, ME	6 hens	20 ft. from property line	$25 fee, enclosed or fenced at all times

Added to the basic regulations are a range of requirements to ensure that the hens will be good neighbors. These can get complex. For example, in St. Paul, Minnesota, prospective flock owners have to get permission from 75 percent of their neighbors within 150 feet of the chicken house, a stipulation that could strain neighborly relations. ("Billy, that's Mr. Grimsley, the man who won't let us have chickens!") Other towns enforce a buffer zone of 100 feet, which seems reasonable until you think of the average lot size. (Chickens only for the rich?) In some cases, chickens are technically prohibited but tolerated unless there is a complaint. In others, the issue is simply not addressed by city codes, leaving a backyard farmer in legal limbo.

Banned in Boston

If your town currently bans chickens, don't give up hope. The tide is with you. There is plenty of precedent for codes that preserve reasonable community concerns (odor, noise, nuisance) while letting responsible flock owners enjoy the benefit of chickens. And precedent is what it is all about. If you can point to the experience of other communities, coupled with the exact language used in their ordinances, you're likely to find your own community much more open to the idea. Should you embark on a campaign to change local zoning regs, be prepared to make an appeal to your city council. The relevant code, if chickens are addressed at all, will fall under zoning jurisdiction or animal control. In rare cases, it may be a public health issue. Here are some of the concerns to expect:

- **Will the chickens cause odor problems?**
- **Will the feed and waste attract rats and mice?**
- **Will the chicken house and pen be an eyesore?**
- **How far from neighbors should the chickens be?**

Most of these concerns can be allayed if the chickens are well cared for. Sometimes, the specifics of good husbandry will have to be spelled out. (See opposite.)

Another concern is how many chickens is enough? Even staunch chicken advocates agree that six chickens—capable of producing almost three dozen eggs a week—is a reasonable limit. Some cities permit as few as three hens. Houston allows up to 30 chickens for showing purposes, but only seven strictly for egg-laying purposes.

Even the tiny backyards of Brooklyn offer enough space for chickens, opposite. Flock owner Paul Baile uses a small chicken tractor to let his chickens graze his garden. (See page 128.)

Space alongside the garage is enough for three or four layers, above. Carving out a small area inside the garage is one of the least expensive ways of sheltering hens.

This setup offers shelter, plenty of scratching space, and easy access for egg gathering—all in an 8 x 8-foot area.

The Milwaukee Proposal

The proposed amendment to Milwaukee's municipal code does a laudable job of affirming backyard flocks while spelling out reasonable requirements designed to ensure that the chickens will be good neighbors.

The proposal begins with a rousing affirmation of the rightness of keeping chickens: "Whereas, the keeping of chickens in the city supports a local, sustainable food system by providing an affordable, nutritious source of protein through fresh eggs. The keeping of chickens also provides free, quality, nitrogen-rich fertilizer; chemical-free pest control; animal companionship and pleasure; weed control; and less noise, mess, and expense than dogs and cats."

Four hens may be kept. Permits cost $10 a year and are granted only when the owner completes "a certification class in chicken keeping from a Master Chicken Keeper to ensure safety and responsibility of the animals and the neighborhood."

Chickens are to be kept in pens but may graze "in a securely fenced yard if supervised. Chickens shall be secured within the henhouse during non-daylight hours."

Neatness Counts. Stored manure must be enclosed and covered with a roof or lid. Chicken coops and pens "must be clean, dry, and odor-free, kept in a neat and sanitary condition at all times, in a manner that will not disturb the use or enjoyment of neighboring lots due to noise, odor, or other adverse impact."

Choosing a Breed

Until diversified farms began to disappear from the scene in the 1960s, chickens were an integral part of farm self-sufficiency. Utility breeds such as White or Barred Plymouth Rock, Rhode Island Red, Buff Orpington, and Wyandotte offered both eggs and meat for the table, and through egg sales, precious cash. Since then, commercial chickens are bred for either egg laying or meat production. However, those early breeds live on today. Red Sex Link, Red Star, Black Star, Rhode Island Red, and Barred Rock are popular brown-egg layers. White Leghorn and California White are productive white-egg layers.

Visiting a county fair is a great way to see different breeds and have a chance to meet owners to learn about a breed's strengths and weaknesses. You may even be able to arrange the purchase of your new flock.

Chicken Profile

Lifespan: 4–7 years, 20 possible
Weight: 5.5 to 9 lbs., depending on breed
Body temperature: 104°F–107°F
Annual egg production: about 265, 300 max.
Feed efficiency: 4 lbs. of feed per dozen eggs
Natural habitat: Asian jungle

A Happy Alternative to Factory Farming

By keeping your own backyard flock, you have a built-in alternative to eggs produced by caged hens confined in beastly conditions. The argument for the factory approach is that it produces cheap eggs. Indeed, animal scientists and egg producers have succeeded in producing an inexpensive protein source in the most efficient manner possible. (Eggs that today cost less than $1.00 a dozen would have cost the equivalent of $4.48 a dozen in 1950, before the advent of caged production.) However, the real costs are hidden. Highly concentrated populations of hens require high levels of antibiotics, a bad habit that may reduce the effectiveness of antibiotics in the future.

Dependence on large-scale operations means that when something goes bad, such as a salmonella outbreak, it goes bad in a big way. And the huge amount of manure produced can be harmful to local air and water quality.

And then there are the chickens themselves. They pay the highest price. Industry guidelines ask for 67 square inches of cage space per bird—less than the size of this page. (Such confinement requires that the end of chickens' beaks be burnt off to eliminate excessive pecking.) Factory farming turns the chicken, a living creature, into a mere unit of production without a chance to scratch in the dust or see the blue sky.

What Breed Is Right for You?

Get some local advice from other flock owners or a regional hatchery. One quick resource is the Web site My Pet Chicken (**www.mypetchicken.com**) whose "Which Chicken" breed selector tool asks you to respond to six questions and then offers you several suitable breeds to consider. The Web site BackYard-Chickens (**www.backyardchickens.com**) has an encyclopedic guide to breeds and their characteristics. You'll find that most chickens are very adaptable and suitable to most climates. With the right housing and care, they'll flourish. A productive layer will give you five eggs a week—even one a day as they reach peak productivity.

Leghorns produce lots of white eggs, take well to human contact, and do well in any climate.

Rhode Island Reds lay up to 300 brown eggs a year. Famous old utility birds, they are intelligent and friendly, and they do well in a wide range of climates.

Sometimes known as the easter-egg bird, the rumpless Araucana and its cousin the Ameraucana produces blue and green eggs. Though only modest producers, they are friendly and take well to confinement.

A great bird for beginners, the Black Australorp is easy to handle and a highly productive brown-egg layer. An all-climate layer, it is also a good meat-type chicken.

The Couchin is one of many breeds worth having just because they're beautiful. This modestly productive layer of light-brown eggs does well in cold climates. She is a calm bird that takes well to people.

Buff Orpingtons are a gentle, hardy breed, able to weather long winters while producing plenty of eggs. Topping out at eight pounds, they are good meat birds as well. Eggs are a pinkish tan.

Structures

A stand-alone coop can be as simple as a plywood box just big enough for your hens, feeders, and nesting boxes. Or it can be a large structure styled to complement your home, replete with window boxes, a cupola, and a weather vane. (See examples on pages 130–133.)

Most flock owners strike a balance in between, but certain requirements are constant for any coop. It needs a good rain-shedding roof and wind-proof walls to cope with rough weather. It must stand above grade to avoid the damp. It needs ventilation year-round (from vents such as those under the eaves in the coop opposite) with additional ventilation for hot weather. (An openable window does the job.)

For the flock owner, storage space as close to the hens as possible cuts chore time, as do built-in nesting boxes positioned at a convenient height and accessible from the outside. The coop should be easy to enter and clean.

If you design and build your own coop, framing with 2×4s is more than adequate. Place wall studs every 24 inches instead of the 16 inches on-center used in home construction. Plan your structure around 4-foot increments to get full use of 4×8 plywood and particleboard panels. Roll roofing is less expensive than shingles and better suited to low pitches. Use pressure-treated lumber anywhere within splashing distance of the ground.

If building from scratch is a daunting prospect, consider a prefab coop. (See page 133.) If you will have only a few hens, you can purchase a ready-made, plug-and-"lay" unit like the Mac-inspired Eglu. (See page 132.)

Coop Specs

- 12-in. x 12-in. x 12-in. nesting boxes
- 8 in. of perch per bird
- 4 in. of feed trough per bird
- 4 sq. ft. of floor space per bird
- 12-in. x 12-in. chicken door

Everything Handy

You'll be checking your hens at least twice a day and cleaning out the coop once a week, so make sure it works for you, too. That means plenty of storage space and easy access for gathering eggs, filling the feeder and waterer, and cleaning out the litter.

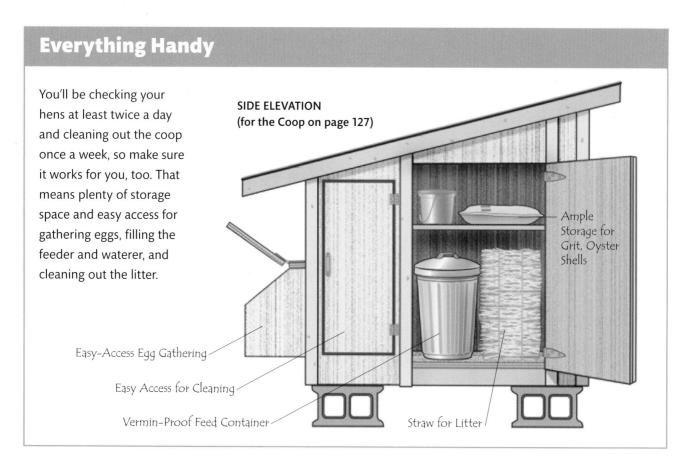

SIDE ELEVATION
(for the Coop on page 127)

Ample Storage for Grit, Oyster Shells

Easy-Access Egg Gathering

Easy Access for Cleaning

Vermin-Proof Feed Container

Straw for Litter

What Chickens Want

A suitable coop can be of any shape or size, but it should have these basic ingredients to keep your hens healthy and happy. You may also want to add electricity for a water heater or for a timer and light to extend laying.

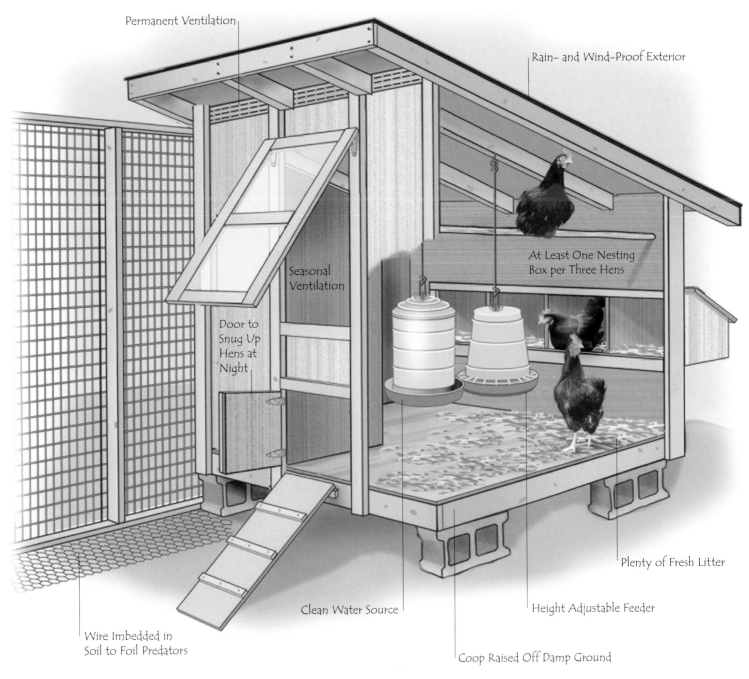

Permanent Ventilation

Rain- and Wind-Proof Exterior

At Least One Nesting Box per Three Hens

Seasonal Ventilation

Door to Snug Up Hens at Night

Plenty of Fresh Litter

Wire Imbedded in Soil to Foil Predators

Clean Water Source

Height Adjustable Feeder

Coop Raised Off Damp Ground

Outdoor Foraging Space

Chicken Tractor

If you like the idea of giving your hens a change of scene while putting them to work on weed and pest removal, consider a chicken tractor. This structure incorporates a pen and coop into a mobile combo that can be rolled or dragged to a new area of pasture each week. Some are only 8 feet long; others can be 12 or more feet long and tall enough to enter. Some flock owners use their tractors as secondary housing, a way of getting the hens out of the main coop for a field trip.

A variation on the chicken tractor theme is a permanent coop combined with a portable pen or two or more pen options surrounding the coop. That way some vegetation will always be available to hens, and the chicken yard won't become a moonscape. Some chicken owners have a portable day pen so that the hens can work over a post-season garden bed or prep new ground for spring planting.

This clever tractor offers housing and even a litter area on the second story and grazing down below.

For Small Getaways

Ideal as a way to let your chickens graze when you don't have fencing, this small tractor has enclosed space for shelter and a waterer but can easily be slid onto fresh grass. Adding shade with a plastic tarp protects the hens on hot days.

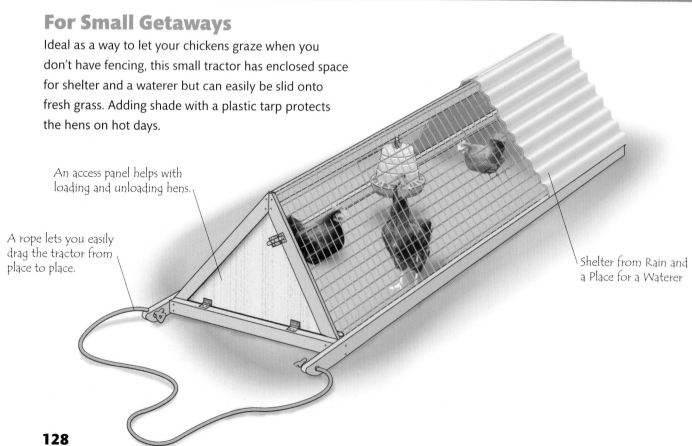

An access panel helps with loading and unloading hens.

A rope lets you easily drag the tractor from place to place.

Shelter from Rain and a Place for a Waterer

A Major Mobile Alternative

A mega tractor like this can be a permanent home for your chickens while you move them weekly to enjoy greener pastures. This version is fully equipped with a nesting box. The challenge with large tractors is building light but strong.

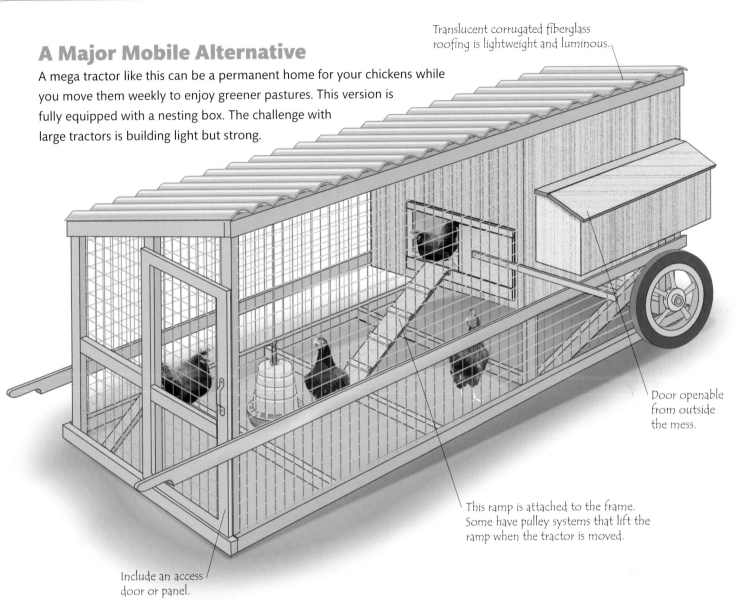

Translucent corrugated fiberglass roofing is lightweight and luminous.

Door openable from outside the mess.

This ramp is attached to the frame. Some have pulley systems that lift the ramp when the tractor is moved.

Include an access door or panel.

Make It from Plans

Designing your own tractor is not difficult, but if you want the benefit of a design with all the bugs worked out of it, consider buying plans. Plans for this versatile tractor are available from Catawba ConvertiCoops **(www.catawbacoops. com)**. For reviews of other available plans visit Chicken Tractor Plans **(www.chickentractorplans.com)**.

Ideas for Coops

Beyond the basics shown on pages 124–125, a good-looking coop is a helpful way to encourage neighbors who may have misgivings about the effect of your flock on property values.

A Cape Cod-style coop, right, looks great and offers plenty of space for man and beast alike.

Chain-link fencing works well for larger runs, below. Topping off the run with mesh protects your flock from birds of prey.

The neat setup at opposite top has a compact scratching area and easy-access nesting boxes—a great solution for a small flock.

Chicken Sublet

The simplest option for housing chickens is "subletting" part of a garage or garden shed. All you need to do is partition off enough space for your hens (allowing at least 4 square feet of space per hen); outfit it with a nesting box, perches, and feeders; and cut a hole leading to an outside pen. This is the cheapest solution and offers superior protection from the weather. It is also handy for humans because of the ease of access and ample space for storing feed and supplies.

More Ideas for Coops

A chicken run doesn't need a lot of height, top left. Placing the coop on legs provides additional scratching space while keeping the overall footprint small.

The crisp little coop opposite offers room to scratch when everything is closed up for the night. It also provides easy egress when it's time to let the hens loose in the yard. A salvaged roof vent formed the basis for the gable roof.

Convenient access to feed and supplies eases chore time, left. This arrangement is well equipped to keep out varmints.

Eglu

The high-tech Eglu, which began as an art school project made by four budding industrial design students, is one of several out-of-the-box options. Pricey but stylish Eglu products offer one of the quickest ways to get started with chickens. They'll set you up with everything you need—even chickens—at **www.omlet.us/ homepage/**.

Prefab Coop

A prefab coop is a great way to quickly house your flock, sometimes costing little more than you'd have to pay for a plan and all the raw materials if you built it from scratch. While it requires assembly, you won't be reinventing the wheel. Check sources **www.Chicken-CoopSource.com** (Chicken Coop Deluxe shown), **www.CoopMyChicken.com,** and **www.horizonstructures.com.**

Getting Started

A laying hen can produce thousands of eggs in her lifetime without ever laying eyes on a member of the opposite sex. A rooster need only be present if you want to produce your own fertilized eggs to raise a flock completely from scratch. If you opt for this approach, bear in mind that any egg you gather from your flock may have a spot of blood in it, the beginning of a chicken embryo. That's something most people don't want to face first thing in the morning.

You have three options for starting a backyard flock: incubating fertilized eggs yourself; buying freshly hatched chicks that you raise to maturity yourself; or purchasing 17- to 20-week-old hens (called pullets) that are ready to lay.

Incubating Your Own Eggs

You can buy fertilized eggs from most hatcheries. The eggs can be shipped, but expect some breakage. If at all possible, pick up the eggs yourself. Also expect to hatch males and females—there is no practical way of knowing the sex of a fertilized egg. Fertilized eggs can be stored for up to seven days at 40–75° F. (Don't store them in a refrigerator. It is too cold for embryos.) Chicken eggs hatch in 21 days. Success depends on maintaining a temperature of 99.5° F. (+/- 2°), having adequate humidity and some ventilation, and turning the eggs. The latter is necessary to keep the embryo from attaching to the eggshell. You can purchase a home incubator that holds up to

Chicks in a Box

Chicks are cute and a whale of a lot of fun for kids. However, chicks take a lot of care and feeding to successfully nurture them to the point—17 to 24 weeks of age—where they lay eggs. And you'll be feeding them for about six months until that first precious egg arrives. Here's the process.

Order chicks online from a hatchery. Chicks cost as little as $2 each when ordered in lots of 25 or more, but when ordering five chicks or less you'll pay $3–$4 each. "Straight run" chicks are a random mix of male and females, fine if you want some cockerels for meat but not so hot if you are after egg layers. Chicks can be sexed with a high degree of accuracy. Expect to pay more for females than males. The newly-hatched chicks are shipped via express mail and will be held at the post office for you to pick up. (How do chicks survive shipping without food or water? They ingest the egg yolk before hatching and can feed off it for 3 days after hatching.) In cold weather some hatcheries add a heat pack to the box. Here's what's needed:

- A warm room with an enclosure to keep out drafts. (An aquarium, cardboard box, or ring of cardboard at least 18 inches tall will do the job.)
- Clean sawdust or wood chips spread on newspaper.
- Some means of warming the area to 95° F. for the first week, reducing by 5° per week until they're at 70° F. at six weeks. This can range from a suspended work light for a few chicks to a 250-watt heat lamp for 50 or so chicks.
- Waterers they can drink from easily but not mess up.
- Feed in a container that limits their scratching it into the litter.

Chicks ship remarkably well because they ingest the egg yolk to nourish them for the first few days after hatching.

Let the Hens Do the Brooding

Letting nature take its course and allowing hens to raise their own brood is an appealing idea. Some flock owners say the resulting chicks are stronger and healthier than those raised by man. After mating, a hen will go broody and stay in the nest while she lays several eggs. She protects her eggs and will almost growl when you attempt to gather them.

Placing a brooding hen in a separate cage with her own feed and water will give her a chance to stretch her legs without interlopers taking over the nest. After 21 days of incubation, the eggs will begin to hatch. It may take a day or two for all the eggs to hatch. The hen, if a good mother, will continue to keep the chicks warm and will introduce them to feed and water.

Not all hens are good mothers, however. Some will give up on brooding after a few weeks, and the embryos will die. Others may even turn on their chicks once they are born.

seven eggs. Some incubators maintain the correct temperature and turn the eggs. Many incubators have settings for hatching duck, pheasant, and quail eggs as well.

Incubating your own eggs is exciting and a great experience for kids but has some downsides. First, you are going to get male chicks. Chicken sexing being tough to learn, you won't know for certain the sex of your chickens until you have expended a lot of feed raising them to 16 weeks of age or more. Get around this by buying only hybrid "sex-link" eggs, which hatch into chicks that are color coded according to their sex. And getting the temperature wrong a few times during incubation could affect their health.

The Ready-to-Lay Option

If you want to skip the brooder stage, purchase ready-to-lay hens, or pullets. Pullets cost $7 or $8 each. Almost every state has a hatchery that sells pullets, though it may entail quite a drive to pick them up. Check the web for private individuals who may have raised more chicks than they need and are willing to sell them. You'll be saving 16–22 pounds of feed per bird by buying pullets instead of chicks—about $14 worth—but you do miss out on a lot of the fun of raising the chicks. Enduring relationships are established in those first few months. You may find a small-animal auction in your area selling a wide range of poultry, often at bargain prices. While many sellers are reputable, you run the risk of buying cockerels mixed in with young hens or bringing parasites and diseases home.

Layers and other poultry can be bought at a small animal auction, often at a savings. However, the health and even the sex of the birds might be open to question. Buy only from sellers with known reputations. Or bring along someone who knows their birds.

Feeding and Watering

In her natural state, a chicken is an omnivorous forager. That's why chickens love a bit of ground to scratch in where they can relish bugs, grubs, vegetable scraps, and weeds. However, working chickens that are laying at peak productivity need supplemental feed that contains protein and carbohydrates, and extra calcium for eggshell development.

Commercial feed is formulated to match the stages of a layer's life. **Starter** gets chicks going in their first three weeks, **grower/developer** fed 6 weeks prior to egg production prepares them, and **layer** maintains them when they are producing eggs. This isn't just merchandising gimmickry. For example, a growing bird will need less calcium than a producing hen because most of the calcium goes into the eggshell. Organic feeds are readily available.

Freshness Matters. Plan your purchasing so that new feed is on the scene every two months. Store it in a vermin-proof container such as a galvanized garbage can well away from moisture, excessive heat, and sunlight. Do not serve stale or moldy feed. If you give the chickens kitchen scraps, give them only enough that can be con-

Can You Mix Your Own Feed?

To produce your own feed you'll need to grow some form of grain (for carbohydrates) and soybeans (for protein). Grinding the harvest and properly combining the ingredients is a lot of work and in the end will likely prove more expensive than buying commercial feed. Most owners put their emphasis on providing a fresh grazing area or raising vegetable crops expressly for their chickens. One chicken owner raises Swiss chard and kale for her three hens but notes that they like organically grown Romaine lettuce the best. Plain old grass is fine as long as it has not been treated with pesticides. If you are raising meat-type birds, buy feed formulated for that purpose.

sumed in half an hour. Even the scraps should be fresh, though chickens dote on sour milk. Chickens that eat onions and garlic will produce redolent eggs.

Washed eggshells are a fine source of calcium, essential for egg laying. Hens love pecking for shells in the litter.

To avoid wasted feed, raise feeders high enough that hens can't scratch the feed out. Chicken feed makes expensive litter.

Are You in It for the Eggs?

Eggs have gotten a mixed rap over the years, but they are undeniably one of the best sources for homegrown protein you can find. Some breeds can produce an egg every 28 hours. If you are in it for eggs, you'll want to know how to boost your productivity and feed efficiency.

Use Nesting Boxes

If chickens had their way, they'd lay their eggs anywhere the spirit moved them. The result is fewer eggs, eggs of unknown freshness, and a free lunch for rats and raccoons. Chickens trained to use nesting boxes will put the eggs where you can gather them daily, not under some handy bush or behind the garden shed. Almost anything can be a suitable nesting box as long as it provides privacy, is about a cubic foot in size, and can hold straw as nesting material. Chickens like to scratch, so a lip on the box keeps the straw from being kicked out.

Increase Light in Winter

Using a light and timer to extend daylight in the wintertime will improve egg production. Daylight, artificial or otherwise, totaling 14 hours a day will keep the eggs coming. Some owners believe this steals from chickens a season of rest they naturally need. Others argue that if humans switch on a light to work on into the evening, why shouldn't their chickens step up to the plate as well?

Good Feed

Eggs demand a lot of protein and calcium (the stuff shells are made of). Regular and adequate chicken feed appropriate to the age of the birds will keep the eggs coming. Supplementing feed with crushed oyster shells or washed and crushed eggshells keeps the chicken from stealing from its bone structure to make eggshells. Commercial chicken feed sometimes combines feed, calcium, and grit.

Cull

Ah, personnel issues. They are the bane of management. Chicken keeping is no exception. Shirkers have to go, or be culled out, if you want cost-efficient egg production. This is especially tough when an otherwise good performer gets old. At best, a layer will taper off in her fourth year and pretty much shut down after seven or eight years. By then she may be an old friend, which is why some flock owners have learned not to name their chickens.

Types of Feed. Commercial feed comes in three forms, **mash,** in which the ingredients are ground and mixed together; **pellets,** which are formed by adding a binder to mash before heating and extruding the feed; and **crumbles,** essentially pelletized feed that has been crushed. Pellets and crumbles are handy if you feed your hens out of doors or are concerned that wind might blow mash feed out of the troughs.

While there is no harm in spreading some feed in an off-season garden to encourage chickens to do some serious scratching and pest removal, place the feed in containers appropriate for the age and size of your flock. Allow for 1 linear inch of feeder space per growing chick and 2–3 inches per layer. An ideal feeder is narrow or protected by a grate so that chickens can't get inside to scratch out your valuable feed. Position the feeder so that as the birds feed their shoulders are higher than their rumps. That will keep them from splashing precious feed. Whatever type of feeder you use, keep it clean. Don't let old, moldy feed accumulate.

Water

Give your chicken constant access to fresh water. A bowl of water is a temporary solution at best; chickens can foul it within minutes and very likely dump it over. A gravity-feed waterer, whether homemade or store bought, is a better solution. You should suspend it above the floor to limit fouling.

True Grit

You've heard the phrase, "Rare as hens' teeth." That's apt because chickens don't have teeth. Instead, they masticate their food in their gizzard by churning it with grit made of pulverized granite or other hard stone. Ranging hens will pick up some grit naturally, but it's a good idea to have grit available to them at all times.

Once chickens start laying they may also need crushed oyster shells to supply the calcium carbonate used to make eggshells—unless the feed already contains adequate calcium. Some feeds contain both grit and calcium.

Feeders don't have to be fancy, they simply need to deliver chicken feed in a way that keeps chickens from scratching it into the litter.

Improvising with plastic containers is a frugal and effective way to deliver clean nutrients. Keeping the containers slightly above the chicken's back limits waste.

Feeders and Waterers

Feeder

Waterer

The do-it-yourself rig at left keeps water available in freezing temperatures. Simply bore four ¼-in. holes along the lip of a plastic tub. Fill the tub with water; set the pan on top; and hold them as you flip the two onto the bricks. An incandescent bulb in the work light below keeps the water from freezing. The feeder and waterer at top and right can hang above the ground to keep the contents clean.

Litter Lowdown

Chickens love to scratch. Litter gives them plenty of working material, as well as being an absorbent that eases cleaning out the coop. For the backyard farmer, another advantage is its usefulness as mulch for the garden. Sawdust, wood shavings, dry leaves, and straw (avoid weed-seed-bearing hay) all work well. Go for the least costly local option.

Sawdust and wood shavings kill odors and do fine on the garden. Avoid sawdust from pressure-treated lumber.

Dry leaves cost nothing and quickly break down in the soil. Laden with droppings they make great fuel for composting.

Chicken Health

Evaluating your flock and removing those hens likely to be poor layers is a process called culling. Even before chickens begin to lay, you should have a good idea of which are the healthy ones and which are not up to par. Those that are crippled or diseased should be removed. Be less quick to remove undersized layers—some are just late developers.

Color is a good guide to health and laying ability. Dull eyes and pale combs and wattles indicate poor health. Intensely yellow eye rings, beak, and feet are a sign that the hen is laying few eggs. Normally, any yellow coloration should be diverted to the egg yolk. Body conformation counts too. A deep, soft abdomen and broad pelvic region indicates good egg laying ability.

Molting

New flock owners should be braced for the annual molt chickens undergo, a rejuvenating process during which chickens stop laying while they replace their feathers. Molting chickens shed feathers from the head first, losing feathers down the body and wings with the tail dropping its feathers last. They don't develop bare spots, however. That is typically the result of being at the bottom of the pecking order. Molting often occurs in the fall, though it can happen any time. An unusual stress can also cause chickens to molt. The complete molt takes about six weeks. One oddity: late molters (12 to 14 months of age) will have a more scraggly appearance yet may be better layers. Early molters are better lookers because they slowly drop feathers over 4 to 6 months, so they

Sick Bay

A sick chicken behaves in much the same way as a sick human. Watch for loss of appetite, listlessness, weight loss, or lack of productivity. These symptoms are warning signs of a condition that may require treatment:

- **Coughing, wheezing, labored breathing**
- **Wound, abscess, warts, swollen joints**
- **Lack of coordination, odd twist of neck, paralysis**
- **Discharge from nose or mouth, diarrhea**

The conundrum is always when to call in the vet. Begin by isolating the ailing bird. Then do a little on-line research. The American Poultry Association's Youth Poultry Club site (**www.apa-abayouthpoultryclub.org**) has a helpful symptom-led site that can direct you to a likely cause. For more detail, turn to the ag departments of the University of Minnesota's (**www.ansci.umn.edu/poultry/resources/diseases.htm**) or the University of Florida's (**http://edis.ifas.ufl.edu/ps044**) site.

Because small flock owners pay close attention to their hens and spare their hens the stress of intensive housing, the hens tend to have few health problems.

never develop the ragged look of quick molters. This means that they will be out of production for a longer time.

If you add artificial lighting to prolong production, the molt may be only partial. Some flock owners reduce light for six weeks during the hens' second year to rejuvenate them.

Temperature Stress

Most breeds turn out pretty tough birds, able to survive wide seasonal variations. It is normal for their body temperatures to rise a bit in hot weather, from 104° to 107° F.

However, no one wants layers collapsing of heat stroke or, at the other end of the temperature extreme, suffering frozen wattles or feet. If you live in a very hot climate, provide your hens plenty of shade. Consider adding misters to cool your flock during intense heat. Cold temperatures require a dry, draft-proof shelter; plenty of clean litter and nesting material; and sometimes insulation as well. If you have one nesting box per hen, consider running a low-watt piano heating tube through the row of boxes. The 25 watts of heat can make all the difference when temperatures drop.

Parasites

Lice. Ticks. Fleas. If you are already scratching involuntarily, you know how hard it can be to deal with these parasites. Refreshing litter often, keeping on top of manure removal, and giving the chickens the chance to dust bathe out of doors lessens the chance of infestation. Still, these pests can attack the best of flocks and test not only your patience but also your desire to avoid toxic insecticides. One prime cause is the introduction of new birds; often they come bearing gifts.

Make it a habit to take a good look at your hens and their environment often. Watch for these signs:

- Small light-brown insects and white stuff stuck to the base of butt feathers indicate lice. Treat with insecticide dip or dust.
- White dust under feeders, in corners, and in other hidden places and tiny, dark-red insects that flee light at night mean you've got mites. Prepare for a massive cleansing of the chicken coop, perhaps including an insecticide bomb. Dip the hens in insecticide.
- A congregation of dark, bouncy insects around hens' eyes spells fleas. Apply baby oil or another mild oily substance to the affected areas. Then cleanse the coop as you would for mites.

Chicken Psychology

Chickens are not always kind. For example, the term pecking order, the domination of the strong over the weak, comes from chickens. Every group establishes a pecking order, and there can be some brutal incidents as a new set of pullets sorts things out. Worse is the late introduction of a new hen to a flock. Hard as it might be, the chickens need to sort this out. Intervene only if a hen is being denied food and water or is injured. Should one hen incur a flesh wound, the other hens will peck at the bloody area, making it worse. This is a hen's least endearing trait. If things get out of hand, isolate the wounded hen until it's healed.

Gathering Eggs

Nesting boxes promote clean, undamaged eggs. Eggs laid on the floor or in the yard risk getting dirty or damaged. Install nesting boxes, and train your hens to use them. Start early. A hen as young as 16 weeks old may start showing nesting behavior—sitting in a corner of coop, pacing with her rear close to ground, or acting like she is searching for something. Pick her up and set her in the nesting box. She'll eventually take to the privacy and cozy nesting material. When the flock reaches 20 weeks of age, add a few fake eggs to the nesting boxes to telegraph the concept.

Gathering eggs first thing in the morning is always good policy to keep the egg from being dirtied and cracked by latecomers. A second gathering in the evening will save eggs from being roosted on overnight. (Expect late-in-the-day eggs to have thinner shells than early arrivals.) In addition, keep the nesting material fresh. Remove any manure that accumulates in the nesting box. Clean out the floor litter often enough to reduce the chance of hens stepping in manure. As you gather eggs, place any that are soiled or damaged in a separate container to avoid contaminating the other eggs.

Going Broody

Hens are wired to hatch chicks and can get broody, settling into a nesting box or anywhere comfortable and refusing to move. If this gets out of hand, do the right thing and lift her out of the box and send her outdoors. She may get aggressive, so you might want to wear gloves when evicting her.

Get in the habit of gathering eggs twice a day so that you'll know exactly when they were layed. You'll also minimize risk of damage when chickens sit on them overnight. A lip on the nesting box helps keep the litter in the box and eliminates the chance that eggs will roll out.

Egg Money

Eggs have long provided family farms with a modest but steady flow of income. Traditionally, a flock of laying hens provided farm wives with rare discretionary spending. As the *American Agriculturist* noted in 1862, "The frugal housewife well knows the advantage of a basket of eggs for the store, to be returned in a few yards of muslin or calico, a spool or two of thread, a packet of tea, and sundry other 'notions'." Well into the 1960s, a flock of 100 or 200 laying hens was a common feature on the farm.

The tradition lives on. Taking a few dozen fresh eggs into work is a sure way to find some enthusiastic customers. The income will defer feed costs a bit, but don't expect a lot of profit. "It is not an economic proposition", says T. J. Johnson, a Washington State backyard farmer who keeps three brown-egg layers. "People do not raise chickens to get cheap eggs. The economy of scale is not such that you can do better than eggs on sale at the supermarket."

The Fresh Test

A fresh egg will sink like a stone when placed in water because its interior is filled with yolk and albumen. As it ages, especially when stored in an arid environment, the interior begins to evaporate and the egg draws in air. If enough air is pulled in, the interior begins to rot, creating even more captive gas. This simple test determines freshness.

A fresh egg sinks. All is well. Your egg is farm fresh.

An older but still acceptable egg will stand up in water. Still good eating.

A rotten egg floats. Handle with care!

Cleaning Eggs

The eggshell is a wondrous thing. Understand it, and you can't go too far wrong in your cleaning and storing technique. The critical thing to know is that although an eggshell appears solid, it actually has 8,000 microscopic pores, to convey oxygen to the chick. In addition, it has a mucous outer coating called the cuticle, or bloom, that is applied just before the egg is laid. This coating prevents bacteria from entering the shell. Overwashing eggs can do more harm than good, achieving a cosmetic cleanliness while actually making the egg more bacterially vulnerable than before it was washed.

For this reason, many flock owners avoid washing all but the dirtiest eggs. Instead, they remove dirt and manure by lightly burnishing the egg with sandpaper or an abrasive pad. This preserves much of the antibacterial bloom and eliminates the possibility of wash water being drawn into the egg.

Some eggs may be so soiled that they require washing. When you have to wash use this technique:

- Spray the eggs with water at least 20° F. warmer than the egg. That typically means water at 90–120° F., hot enough to warm the contents of the eggs slightly to create an interior pressure that keeps contamination out.
- Use a nonfoaming unscented laundry or dishwashing detergent.
- Finish the washing with a dip solution made of one tablespoon of bleach to a gallon of hot water.
- Give the egg a final warm rinse.

Sanding off bits of dirt is preferable to washing because the protective membrane on the outside of the shell is preserved.

Storage

Store eggs by placing them in a carton marked with the date they were laid. Eggs can be safely stored in a refrigerator for a month or more. (In fact, in cool weather you may find eggs can sit on the kitchen counter for weeks and still stay fresh.) Many people successfully store eggs in a basement or root cellar for months. If you attempt long-term storage, leave them unwashed to preserve the protective bloom. If you have doubts about the freshness of an egg, gently place it in a glass of water. (See box, page 143.) If it floats, the interior has begun to evaporate, drawing in contaminated air. Discard the egg. Candling eggs is used to spot fertile eggs but can also reveal an egg that is past its prime. Candling involves holding an egg in front of a bright light and judging the extent of the air space. More than ⅛ inch indicates an iffy egg.

Eggs for Sale

Before selling surplus eggs, check your state's regulations. For example, North Carolina requires that anyone selling more than 30 dozen eggs per week grade them and indicate the grade on the carton. Farmers selling fewer than 30 dozen eggs per week must use clean (not necessarily new) cartons with their name and address and the word "Ungraded" on the carton. The eggs must be clean (washing not required) and immediately stored at 45° F. Your state may have different requirements. Poultry supply and hatchery sites sell egg cartons in lots as small as 50, which would be more than enough to get started. Or you can simply ask friends to recycle cartons in your direction. Either way, your customers will likely be happy to bring in an old carton when they buy eggs.

Meat-Type Chickens

While most municipalities prohibit chickens raised exclusively for meat, meat-type breeds, if allowed, grow quickly with minimal fuss. If you are raising exclusively meat-type birds as opposed to dual-purpose utility birds, purchase feed formulated accordingly. A Cornish/White Rock cross, for example, will grow to 5 or 6 pounds in 8 weeks. Both males and females are suitable for the table. Exclusively meat-type

When storing eggs, mark the carton or container with the date the eggs were gathered so that there is no confusion later about which eggs to use first.

birds are hybrid combinations of established breeds like the Plymouth Rock or Wyandotte. The resulting hybrid vigor produces the hardiness and quick growth that makes a desirable bird. As they grow to full weight, meat-type birds become overcome by their size and are less active than layers.

(The Cornish Hen is not a separate breed but merely a bird with Cornish parentage that has been butchered young—usually at 5–6 weeks of age and 2 pounds of weight.)

Chickens bred for meat alone put on weight fast, growing to 5 or 6 lbs. in 8 weeks. Cornish/White Rock hybrids, right, are a popular choice.

Killing and Butchering

Anyone who regularly butchers chickens has honed an approach that gets the job done quickly and with minimal angst. What follows is an overview of a couple of methods. Killing and butchering warrants a good mentor who can demonstrate the techniques firsthand. If you plan to butcher your chickens, find an experienced practitioner, and help out a few times before going solo.

The process begins well before the killing. A day in advance, separate the chickens you plan to butcher. Keep them away from food to allow their systems to clear out. As to the killing itself, the old-fashioned chopping-block method is dangerous, messy for executor and executee alike, and unnecessarily traumatic. It is the source of the phrase, "Running around like a chicken with its head cut off," for some chickens will actually do so—a sight you won't soon forget. Another old-fashioned method is to grasp the chicken by the head and whirl the body around to twist the neck, effective but not for the fainthearted. Instead, use one of these methods, each of which involves holding the chicken upside down, a position that calms and befuddles it.

For a less traumatic option, hold both feet of the chicken by one hand and let the chicken hang upside down until it is quiet. Hook the back of the head between your thumb and forefinger, and push down to break the chicken's neck. Immediately cut its throat to bleed it. Or gently place the chicken upside down in a sheet metal cone made so that the head will protrude. Use a very sharp blade to quickly sever the arteries on either side of the neck. The chicken will bleed to death quickly.

Once the chicken is bled, pluck the feathers. You can do so by hand, grabbing enough feathers that the job goes quickly but not so many that you tear the skin. Dipping the chicken in hot water releases the feathers, so they are easier to pluck, but is uncomfortable and about as foul smelling a task as you'll ever attempt. If very fine feathers remain, singe them off using a propane torch.

Gutting a bird is something you may want to watch demonstrated before you go solo. There are several methods. One involves tying the anus shut to avoid contamination, then cutting around it to release the intestines from the rear of the bird. Do not pierce the intestines. Cut off the head at the base of the neck. Reach in and loosen the internal organs from the inside of the body cavity. Then firmly pull on the anus end to remove the organs. Check that you've done a complete job of it. Cut off the feet and thoroughly wash the bird before freezing it.

Ducks

Quick-growing and hardy, ducks are less aggressive than chickens and sometimes used to calm skittish laying flocks. While they can be noisy when roused, ducks suit backyard farming well. They love to forage and relish such garden pests as slugs and snails. They are social creatures and are quick to bond with their owners.

Duck eggs brood more easily than chickens, needing only a 250-watt brooding lamp in a sheltered enclosure. Raise or lower the lamp so the center of the brooding enclosure is 98° F, the outer edge 90°. Lower the temperature 5° per week to reach 70° at six weeks of age. Allow about 6 square inches per duckling (1 square foot after a couple of weeks), and lay down 4 inches of wood shavings, chopped straw, or peat moss for litter. Clear out damp areas of litter. Put out fresh water in containers with wire guards so that the ducklings will stay dry—

important when they are at the down stage. Ducklings thrive on chick starter and grow quickly, some breeds reaching maturity at six weeks.

While famous for their meat, some breeds of ducks are surprisingly good egg layers. For example, the Khaki Campbell duck can produce 250 to 340 eggs a year. Duck eggs are bit richer than chicken eggs, but otherwise identical in color and texture. Many people prefer duck eggs to chicken eggs. A meat breed like the Pekin are mature at 9 weeks of age, weighing between 6 and 7 pounds. Other popular breeds are the Rouen and Muscovy.

Ducks can be set up with essentially the same housing and fencing as chickens. The only difference is that ducks prefer nesting boxes at floor level, not elevated. They also welcome a pond but don't require it—they'll happily get by on a bucket of clean water for head dipping. Ducks like company, so plan on at least three. To limit waste, give them crumble or pelletized feed.

The quintessential duck, the Pekin is hardy and friendly—a good starter duck. Bred for both meat and eggs, a Pekin can reach market weight (7 lbs.) in 9 weeks and can lay up to 145 eggs per year.

Muscovy ducks are personable and quackless—a neighborly trait. Their meat is leaner than that of other ducks. Males reach 12 lbs. Females may fly but don't fly away from their familiar roost.

Geese

Beautiful and protective (Romans famously used them as "guard dogs"), geese are hardy and easy to raise. They are an ample source of meat, and even one egg can make a hefty omelet. Hobbyists prize their eggs for craft projects. Though they can be fiercely protective, they are sociable animals that fit in well with other backyard stock like chickens, ducks, and goats.

Brood geese much as you would ducks, using a 250-watt brooding lamp in a sheltered enclosure, 4 inches of litter, and plenty of chick starter and fresh water available. At four weeks or so, change over to chick grower. Geese reach maturity at about 6 weeks of age.

Mature geese love to graze on grass and are good at clearing up a garden at the end of the season. Supplement their foraging with chicken feed and a source of grit. An acre of grass will support about 35 geese. They need shelter from rough weather with plenty of straw. Like ducks, they like a tub of clean water for dipping their heads.

Popular breeds included the dark gray Toulouse, a prolific egg layer as well as a good meat source. Females reach 20 pounds, males 26 pounds. The white-feathered Emden hits about the same adult weights. The African breed is brown feathered with a distinctive knob on its head. It too is a productive layer, but smaller in size with females reaching 18 pounds and males 20 pounds.

Market weight for a goose is 18–19 pounds. Its meat is dark and more flavorful than that of turkey. It is also more moist, thanks to its inherent fat content. Goose fat is treasured by connoisseurs as a rich cooking medium. The fat is removed by pricking the carcass and blanching with hot water or by slow roasting for about four hours, spooning away the fat as the goose cooks—techniques that work equally well for duck.

The Toulouse is an attractive, fast growing goose. A goose reaches 20 lbs., a gander grows to 26 lbs. Seldom a wanderer, this breed stays close to its feed and has a placid temperament.

The Emden gander grows to 30 lbs., the goose to 20 lbs. Females lay up to 40 eggs a year.

Turkeys

Unlike chickens, ducks, and geese, which flourish when kept together, turkeys are a breed apart. While beautiful, curious, and affectionate, they are tricky to raise and are vulnerable to the blackhead organism, a disease that chickens carry but don't fall prey to. That means that unless you have enough space to completely isolate your turkeys from your chickens, choose one or the other.

Experienced turkey owners ruefully proclaim that turkeys seem to look for an opportunity to die. For example, during brooding they have to be trained to eat the right thing lest they gorge themselves on litter and starve as a consequence. One of the simplest tactics for heading them in the right direction is to cover the litter with cloth or coarse paper and set multiple waterers and feeders about for the first week or so. As the birds mature, use only waterers with wire guards—a bucket of water can tempt a turkey, out of curiosity, to plunge its head in the water and drown.

That said, turkey raising has its benefits. Many small-scale turkey farmers are doing us all a service by raising breeds different from the limited number of breeds factory-scale turkey operations use. Commercial turkeys, bred for plenty of white meat and quick growth, not only cannot breed but can sometimes hardly stand. Small farmers maintain genetic diversity by raising breeds like the Standard Bronze, Bourbon Red, Narragansett, Jersey Buff, Slate, Black Spanish, and White Holland. They've done so by the counterintuitive strategy of encouraging people to eat such turkeys. Known collectively as heritage breeds, these turkeys have found a niche market among people willing to pay a premium to eat birds similar to those our ancestors did. The higher cost is due to the longer growing time heritage breeds require, but many people swear by the richer flavor. Turkey raising might be right for you if you have plenty of land, are willing to do without chickens, and have an entrepreneurial bent.

For plenty of white meat, standard heavy white turkeys can grow up to 45 lbs. if you let them. A typical growing period is 24 weeks to reach about 20 lbs.

Heritage breeds, like this Narragansett, yield dark, flavorful meat. These old breeds are being preserved because more and more "niche-market" customers choose to serve them rather than grocery store loss leaders, for Thanksgiving dinner.

Quail

Weighing in at about 7 ounces (10 ounces for some crosses), quail are ideally suited for small enclosures. They lay exquisite ⅓-ounce eggs prized for their delicate flavor and their acceptability to people otherwise allergic to chicken eggs. Quail also deliver lean meat with a distinctively wild flavor.

Coturnix quail (also know as Japanese quail) and Bobwhite quail are the most popular breeds and the nearest thing to a domesticated quail. In fact, quail owners consider them even more friendly than chickens. Because quail rarely breed in captivity, most flock owners start with fertilized eggs from their own flock or from breeders and incubate them until they hatch 17 days later. The chicks are brooded much like chickens. They will begin laying as early as 7 weeks and can lay between 200 and 300 quarter-size eggs per year. If raised as meat birds, they'll reach full size at 7 weeks. Two birds make about one serving.

In most states quail are considered game birds and may require a permit if you plan to raise them.

Coturnix quail like these can be raised exclusively in cages but love to graze. In the summer, they like to take two or three dust baths each day.

Considering Emus?

With their dark, beeflike meat and fat that produces an oil said to have many beneficial properties, emus are an intriguing livestock option for small farms. Here's what it takes.

Emus eat grass, grain, bugs, worms, and some vegetables and fruit. They lay avocado-size eggs. They are ready for market in 18–20 months but can live for up to 30 years, with females laying eggs until they are 15 years old.

Emus grow to 5 feet 6 inches tall and weigh in at 150 pounds. They are strong runners, hitting speeds up to 40 mph, and must be penned behind 6-foot fences. Growers keep them in 25- x 100-foot runs behind heavy gauge wire fencing. Females are more aggressive than males. (The males are of a nurturing persuasion and are the ones who hatch the eggs.)

All this adds up to trouble for the small-scale farmer. In-town raising is out of the question—emus are simply too large and too strong. Add to that a volatile market for their meat and their incompatibility with other animals, and you might want to leave emus to braver souls.

5

Raising Goats

Are Goats Right for You?

PERSONABLE, PASTURABLE, PRODUCTIVE—what's not to like about goats? They are tidy, fastidious in their eating habits, and just plain fun to have around. In fact, with more people on the planet drinking goats' milk than cows' milk, it is a wonder goats have not caught on more in North America.

All that is changing. Cities like Portland and Seattle allow miniature goats, and scores of smaller towns permit people to keep goats. And goat products are catching on more than ever. Feta cheese, made from goats' milk, is universally popular. The health benefits of goats' milk are widely touted. And with an ever-growing Hispanic population (among many other goat-favoring people) in the U.S., demand for goat meat is on the rise.

For many people, the strongest argument in favor of goats is how lovable they are. "Goats just naturally gravitate to people," says Phoebe Larson, who keeps a flock of goats near Rockford, Illinois. She emphasizes that goats are social animals. They have to be kept with other goats and have been known to slowly pine away if they lose a companion. And they are clean, especially when it comes to food. "Goats are finicky eaters," adds Larson. "They won't eat anything dirty. That's why the stereotype about them eating garbage is so wrong." They are curious, love to climb, hate getting wet, relish the taste of a freshly fallen autumn leaf, and have a deerlike grace.

Goats are naturally inquisitive and seem to just radiate intelligence, opposite. They make great companions in your farmsteading venture, right.

There is a price to pay for all that goats have to offer. Goats are a serious livestock commitment, requiring a minimum of twice daily attention, at least 15 square feet of shelter per goat and about one-quarter acre of pasture for 3 to 5 goats with a 4- to 5-high-foot fence around it. They need a roofed area to keep them out of the rain and shelter from the wind. (See page 152.) You should be willing to play vet when it comes time to trim their hooves, worm them, and help with birthing. And you need a support group. "It's a really, really good idea to find somebody who has been raising goats for years," suggests Larson. She points out that vets often give larger livestock priority and may not be as well informed about goats as they should be.

Goat Profile

Gestation period: 5 months

Recommended breeding age: 1 year

Number of offspring: Typically 2, but up to 5 is possible

Productive milking life: 10 years

Typical life span: Does 12 years, wethers (castrated males) 14 years, bucks 9 years.

Adult weight: Pygmy goat, 50–75 pounds; dairy goat doe 125–200 pounds.

Body temperature: 102.5° F to 104° F

Food: Hay, supplemented with grain

Quarters for Goats

Goats are hardy and adapt well to temperature extremes as long as they can keep out of the snow, wind, and rain. They should have a yard with a roof over it so that they can be out of doors but sheltered from the rain. As long as they have a three-sided shelter with plenty of straw litter, they should never need to be confined, even in the roughest weather. Many keepers of full-size goats like 5-foot-high fences to keep them from jumping out of the pen; keepers of pygmy goats speak of their gift for getting under fences. Woven wire, plank fences with wire along the bottom, five-strand electrical fences, and portable electric-net fencing are all workable options.

Good fences, right, make for happy neighbors: a goat's love of browsing can be the death knell for nearby ornamentals and fruit trees. Goats are as likely to crawl under a fence as jump over it. Climbing structures, below, engage active minds and bodies.

What Goats Want

Above all, goats like to be dry and out of the rain. However, they are primarily outdoor creatures and always want the option to head out when the mood hits them. Even in the winter, they need only a three-sided structure as a wind break and plenty of litter. They are fastidious and like lots of clean litter.

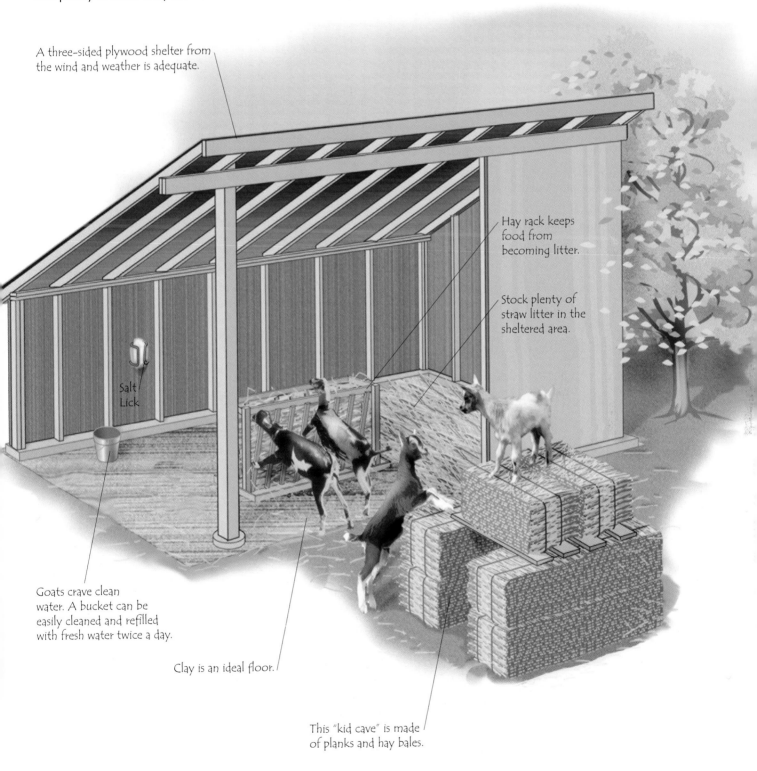

A three-sided plywood shelter from the wind and weather is adequate.

Salt Lick

Hay rack keeps food from becoming litter.

Stock plenty of straw litter in the sheltered area.

Goats crave clean water. A bucket can be easily cleaned and refilled with fresh water twice a day.

Clay is an ideal floor.

This "kid cave" is made of planks and hay bales.

Goats are browsers, not grazers like sheep, and love to reach up to get their meal. Hay is their primary food, but if it falls into the mud, goats consider it litter and won't eat it. Even a favorite treat like an apple core or some tortilla chips will be ignored if dropped in the dirt. That means that some sort of hay rack for keeping hay off the ground is necessary. A bucket of fresh water, cleaned and replenished every day, is the best water source. Their minimal grain ration—just a cup or two a day—can be fed out of a pan or a bowl attached to their milking stand. Buy a plastic or metal garbage can to store grain; keep hay and straw litter out of the weather. Warning: goats are ingenious at stealing food and can contract overeating diseases if they consume too much grain. Store feed behind a latchable door.

You can either purchase a ready-made milking stand or build one. (See page 164.) While most goats are quickly trained to jump up on the stand, a ramp may be needed if the goat is small or the stand is too high.

If you already have sheep, ponies, geese, or chickens, rest assured that your goats will be friendly neighbors and mix right in. However, they need some sort of climbing structure and have been known to climb up onto low tree boughs if the mood strikes them.

Goats love their food, especially all-stock sweet feed and other treats. Hay can be eaten on demand, but grains should be kept out of reach. That may mean storing food behind bars because goats can be ingenious at getting something they want.

Hay vs Straw

Straw is made of stalks left over from the harvest of wheat, oats, or barley. It makes ideal litter. When added to your garden as manure-enriched mulch, it won't bring a bunch of weed seeds with it. Hay is more variable stuff. Sometimes it is merely cut pasture grass, nutritious but laden with weed seeds; a legume crop like alfalfa or clover; or grasses such as orchard grass, timothy, or fescue.

STRAW

HAY

Prefab Alternative

Goats don't need elaborate quarters, but they do need shelter. If you are not comfortable swinging a hammer, here are some options. This trim little shed includes a sleeping shelf that goats like to sleep both on and under. This one is available from Eagle Shed **(www.eagleshedsandgazebos.com).** Small pole barns work well, too, though be sure to line the interior walls with plywood—common aluminum sheathing dents easily.

With plenty of hay storage overhead, these goat shed stalls are ideal for birthing and caring for young kids.

Assessing the Breeds

Reading up on goats will give you a good idea of the right breed for you, but before making your final decision, visit some goat farmers to get the real low down. The fundamental question is what you want goats for: milk, meat, or just the delight of having them around. Breeders tend to be passionate about their chosen goat and may be more tolerant of their foibles than you are. For example, Swiss goats are beautiful, productive goats but difficult to breed except in the fall. Nubians, on the other hand, breed year-round—an important trait if you want a steady supply of milk. (See page 163.)

Popular dairy breeds include the Nubian, Oberhasli, LaMancha, Saanen, and Toggenburg. Meat types include the Boer, Kiko, Myotonic, and Spanish. Angoras are famous for their fiber, mohair.

Dairy Goat Breeds

Famous for its long, floppy ears, the Nubian is one of the oldest breeds and the most popular in the U.S. It gives plenty of high butterfat-content milk.

The medium-size Oberhasli is one of the oldest established breeds, originating in Switzerland.

Second only to the Nubian in popularity in the U.S., the Saanen is large, topping out at about 135 lbs. It is a commensurately large milk producer.

Nicknamed "Toggs," Toggenburgs prefer cooler weather and are known to be spirited. They are moderate-size, and medium-range producers.

Miniature Goats

Pygmy goats, among other miniatures, are ideal for the backyard farmer—and often the only type of goat permitted by cities. Pygmy goats originated in West Africa. Intelligent, good-natured, and affectionate, they make great pets. "Kids are more attracted to them because they are not as intimidating as a regular dairy goat," says Dori Lowell of the National Pygmy Goat Association. She adds that their "teddy bear look" doesn't hurt either. In addition to their good looks and winning personality, they are also steady milkers, producing a pint or two a day.

While not as numerous in the U.S. as Pygmy goats, Nigerian and Nubian Dwarf goats are growing in popularity and are also very suitable for small farms. Their body conformation is more proportionate than that of Pygmy goats, so they truly look like miniature goats. Does often give birth to three kids at a time—sometimes as many as five!

Male Pygmy goats have beards and sometimes dramatically draping manes. They are strong and not recommended as pets. Never let children play butting games with a Pygmy goat, no matter how cute they look.

Pygmy goat colors include white, caramel, black, and two-tone combos. "Agouti" blends of brown and white and black and white are especially attractive.

Pygmy goats are precocious breeders, producing 1–4 kids in 9–12 months. Kids are weaned by 3 months of age and are about as cute as anything in the animal kingdom.

Meat Breeds

We are late to discover goat meat, one of the most widely consumed meats in the world. Goat meat, also known as chevron or cabrito, requires less land and less feed to produce than beef. It's lower in cholesterol than chicken and higher in protein than beef, and long has figured in Middle Eastern, African, Mexican, and Jamaican cooking. The USDA reckons that goat-meat production is growing by 10–15 percent per year due primarily to changing demographics.

Meat breeds mature more quickly than dairy breeds and put on more weight as they do so. While any goat is a potential meat producer, meat-type goats excel at adaptability to their environment, fecundity, growth rate, and carcass value. Meat breeds aren't ideal for small setups, but they are certainly worth considering if you have an acre or two.

The Boer, first bred in South Africa, was developed for shape, high growth rate, and fertility. It is the most popular meat-type breed in the U.S.

The Spanish breed descended from goats brought to the Southwest by Spanish explorers. Until the arrival of the Boer, the Spanish was the predominant meat-type breed.

Myotonic goats are hardy and fertile, and they have a body shape that is right for meat production. They're also known as Tennessee Fainting goats because of their tendency to stiffen and fall over when frightened.

Bred in New Zealand, the Kiko is large framed and typically white. Known for its ability to thrive in a wide range of climates and less than ideal field conditions, it is hardy and productive.

Choosing Well

Reputable goat breeders are a good source of healthy stock, as are dairy-herd owners selling kids after does have freshened, or begun lactating. If you have any active 4-H clubs nearby, check for participants with goat projects who might have animals to sell. Small animal auctions can offer bargains if you choose carefully and can trust the seller. If possible, have a trusted mentor—someone who knows goats—help with the choice. Here's what to look for:

- **Avoid a goat whose head is down, has its tail tucked in, or has a messy behind. Other signs of poor health are swollen joints or a patchy, scruffy-looking coat.**
- **Check for trimmed hooves, a sign of good care. (See page 167.)**
- **Conformation—general body shape and posture—is an indication of fitness. Look for feet set squarely beneath the animal, wide-set front legs, a long and trim neck, and a level rump.**
- **A kid who has already been disbudded—the horns have been removed—will save you an unpleasant but necessary chore. (See page 166.)**

Dealing directly with a reputable local dealer is probably your best bet for buying a satisfactory animal, but always check hooves, health, and general demeanor. It helps to have a mentor along the first time you purchase goats.

Feed and Water

In the wild, goats are browsers, living well off brush and leaves. They have four stomachs that break down and ferment vegetation, creating microorganisms that produce the proteins that actually nourish the goat. While dairy goats need a grain supplement and any goat benefits from a little grain, their basic diet is hay and grass. "Generally, people overfeed their animals," says Chuck Bauer, who has kept goats on his Batavia, Illinois, farmette for more than 20 years. "Unlike pigs, goats can literally eat themselves to death. It is better to err on the side of too little than too much." Indeed, one of the few maladies goats fall prey to is Enterotoxemia, often called overeating disease. Should a goat eat too much grain, normally beneficial bacteria can explode exponentially and produce a toxin that can lead to profuse diarrhea in an adult and can kill a young animal.

Sadly, it is often the grain treats goats love so much that lead to this condition. Vaccination can treat Enterotoxemia, but the right feeding regime is the best preventative. Some owners offer so much grain that goats lose their taste for their natural feedstuff. "Never let it get to the point where they won't eat hay," Bauer asserts. He suggests cutting in half the recommendations on the feed bag.

The essentials are hay morning and night with about a cup of grain for nondairy goats, dairy pygmies, and dwarfs. Give full-size dairy goats 1½ cups of grain morning and night. (All-stock sweet feed is a common choice.) If you must change feed, do so gradually, adding in a bit more of new feed day by day until the changeover is complete. Treats such as fruit, watermelon, squash, greens, and snack chips are great motivators but should be fed sparingly. Always have fresh water available in a clean container. Goats also need a wall-mounted salt block.

A little all-stock sweet feed goes a long way with goats. Dairy goats need only 1½ cups at each milking; nonmilkers should get even less.

A salt block should be always available. And like everything else fed to goats, it needs to be off the ground.

Do Goats Eat Tin Cans?

Old animated cartoons and comic books portray goats as being omnivorous to a fault, happily devouring boots, long underwear, and tin cans. In fact, goats are highly fastidious eaters and would never touch such junk. Except for those tin cans. "What they are looking for is that paper on the can," explains Dori Lowell of the American Pygmy Goat Association. "They like to eat paper. You never want to leave a paperback book sitting next to a goat because they're going to grab it. And they are going to eat it."

Goats are browsers more than grazers, preferring to reach up for food, right. You can spot a goat pasture from the highway by the trees and bushes neatly cropped 5 feet from the ground.

Autumn leaves are a delicacy, just about the only thing goats will eat off the ground, below. Some owners save leaves as a wintertime treat.

Breeding and Birthing

At the very least, a doe will come into heat in the fall, but many breeds can mate year-round. If you don't keep a buck, board your doe with a breeder while mating. Or you can artificially inseminate—a do-it-yourself job once you've purchased some equipment and learned the technique.

Goats are hardy birthers. Many goat owners speak of checking a pregnant doe one night and finding a healthy pair of kids up and running in the morning. Unassisted births are by far the norm, with both front feet coming out first, followed by the head. However, sometimes the doe will need help because the kid is positioned rump first, or the head or one foot may be bent back. A baby monitor is a handy way to keep tabs on an imminent birth. If things don't start moving for a doe within a hour of when her water breaks, she may need assistance.

Nursing

Newborn kids need the benefit of their mother's milk for the first few days. In fact, they should suckle 10 to 20 percent of their body weight in colostrum in the first 24 hours. Keeping the doe and kids in a small pen for the first few days encourages this. When to wean them to the bottle is a choice you'll have to make. "I leave them with their mothers until at least 12 weeks old," says Maggie Leman, an experienced Pygmy goat breeder in Durham, North Carolina. "I don't believe in weaning them earlier. I give them the best start possible." Other owners feel that kids do fine when bottle fed and become used to human contact quicker.

The buck is the source of goats' reputation as a smelly animal. Does are virtually odorless. Bucks can be mean and will need a separate pasture when they are not on duty.

A young kid is just magical, a high point in the cycle of keeping goats. "They are so exuberant, so full of life," says Phoebe Larson, who has kept goats for 30 years.

Goat Talk

Doe: female goat, of any age

Buck: male goat of any age

Kid: young goat of any gender, sometimes used in combination with gender; doe kid, buck kid; doeling, buckling.

Wether: buck that has been neutered.

Group of goats: herd, flock, tribe, drove, or trip

Milking

A doe starts producing milk only after she mates with a buck and gives birth (typically to twins) five months later. Milk production will peak for the first two months—the better to get her kids started—and then level out for the next seven months or so until it is time for her to "freshen"—to again bear kids to restart her milk production. A doe can get pregnant at a frightfully young age, a good reason to separate doelings and bucklings after a couple of months. Waiting until a doe is at least one year old ensures a healthy delivery for the mother and children. A doe typically bears two kids, but as few as one and as many as five are possible.

The Ever-So-Handy Milking Stand. Milking a goat is made much easier with a milking stand. Numerous commercially produced types are available, though it is easy enough to make your own. (See illustration, page 164.) Goats quickly learn to jump up on the stand. Goat owners with a stand for each goat say that each doe will select a favorite stand and jump right up on it at milking time. Food motivates goats and a cup or two of grain (often that beloved "all-stock" molasses flavored feed) poured into the bowl of the stand will bring the does running.

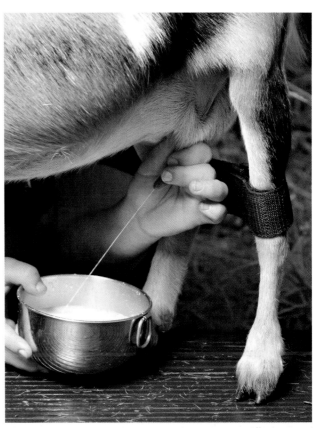

With a little practice, good milking technique becomes almost second nature. Affordable mechanized milkers even are available for small herds.

Milking Technique

Cows have four teats. Goats have two teats, ideal for two-handed milking. ("We like to say that if God intended us to milk cows he would have given us four hands," jokes one dairy goat owner.) The milking itself involves using your thumb and forefinger to squeeze the top of the teat, then using your other fingers to strip the milk out. The technique quickly becomes second nature. Pygmy or dwarf goats have smaller teats that need only three instead of four fingers to strip the milk from the teat.

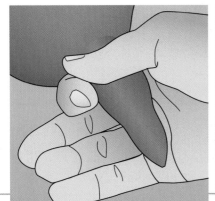
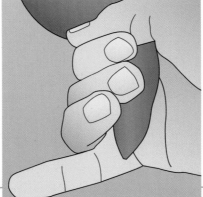
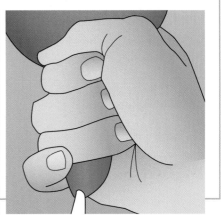

5 Raising Goats

To Wash or Not to Wash

Some dairy herd owners wash the teat before milking, but many don't, arguing that goats, unlike milk cows, are very clean (you'll never find a smear of manure on a goat flank) and the warm water used to wash the teat introduces more bacteria than it removes. "I rub my arm back and forth under [my doe's] belly to knock off any loose hair or chaff," says Larson. "I don't wash them or dip them afterwards." Gently nudging and rubbing the udder like a kid would helps bring down the milk.

Full-size goats need milking twice a day, with the milkings 12 hours apart. One dairy herd owner prefers milkings at noon and midnight, but most prefer a morning-evening schedule. Pygmy or dwarf goats need only be milked once a day—a bonus. If you find these goats just too small to milk easily, consider buying a mechanized milkers based on breast-pump technology. Milk should be refrigerated. Excess milk can be frozen or made into cheese. (See "Maggie Leman's Feta Cheese Recipe," opposite).

Goats quickly take to the milking, not least of all because they are highly food motivated and look forward to their twice-daily ration of grain.

Build-It-Yourself Milking Stand

Simple enough for even the moderately handy, this milking stand can be made in a day using primarily ¾-inch plywood and 2x4s and 2x3s. Choose a height from which it is convenient for you to milk (16 inches being typical), and add a ramp for the goat if necessary. Overall height is about 52 inches.

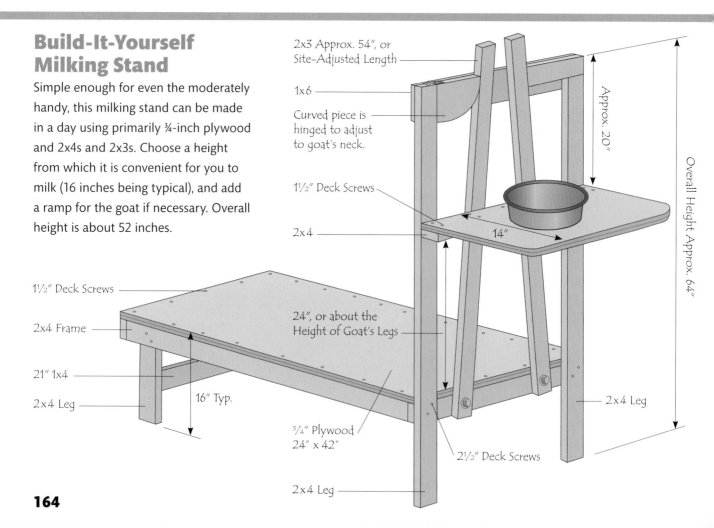

2x3 Approx. 54", or Site-Adjusted Length

1x6

Curved piece is hinged to adjust to goat's neck.

1½" Deck Screws

2x4

14"

Approx. 20"

Overall Height Approx. 64"

1½" Deck Screws

2x4 Frame

21" 1x4

2x4 Leg

16" Typ.

24", or about the Height of Goat's Legs

¾" Plywood 24" x 42"

2½" Deck Screws

2x4 Leg

2x4 Leg

164

Keep Milk Flowing

The tricky thing about milk production is scheduling breeding so that does are freshened in a sequence that ensures a steady milk supply. Each doe produces a reasonably steady supply of milk for 10 months. She then needs to be dry for about two months before giving birth—a pause to focus her energy on the expected kids. That means breeding about 7 months into milk production so that 5 months later the doe gives birth and can again be milked. If you have three does, breeding them in the staggered fashion shown on the time line below produces a steady milk supply with at least two does always producing.

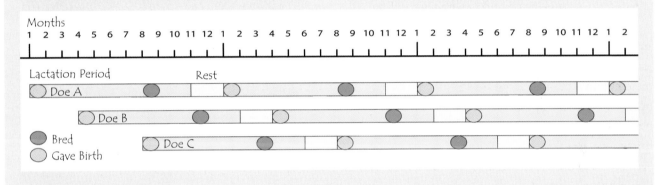

Maggie Leman's Feta Cheese Recipe

1 gallon of goat's milk, held at 86° F.
1 packet of direct set mesophilic starter culture
    or 2 ounces of homemade mesophilic starter
    culture or 2 ounces of buttermilk.
¼ tablet rennet or ½ teaspoon liquid rennet

Mix the mesophilic starter or buttermilk with the goat's milk. Ripen for 1 hour, holding it at 86° F. Dissolve the rennet in ¼ cup cooled boiled water. Gently stir the rennet solution into the ripened milk for several minutes. Cover, and let it set 1 hour, holding the temperature at 86° F. Test the curd. If it breaks cleanly over an inserted finger or thermometer, it is done.

Cut the curd into ¼-inch cubes. Allow the curds to rest for 10 minutes. Then gently stir the curds for 20 minutes. Pour the curds into a cloth-lined colander; tie the corners; and hang for about 12 hours until the cheese is quite firm but not overly dry. This lets the cheese become acidic enough to stand up to being brined or preserved in olive oil. From here:

1. You can cut the cheese into ½-inch cubes, lightly salt it, and store it in the fridge for up to a week.

2. Cut the cheese into ¾- to 1-inch slices, and age in a brine solution for up to a year. To make the brine, dissolve 10 tablespoons of Kosher salt in 5 cups water.

3. Preserve the feta in olive oil and herbs. Maggie has feta that has aged for two years.

Cut the cheese into 1-inch cubes, and salt it lightly for 4 to 8 hours. Into sterilized half-pint canning jars place 1 to 2 cloves of crushed garlic, 5 to 10 whole black peppercorns, and a blend of basil, oregano, and thyme. (Sterilize fresh herbs for 1 minute in the microwave.) Fill cheese cubes up to the shoulder of each jar. Fill with olive oil, making sure all air bubbles are released and the cheese is below the oil. Store in the fridge. When it comes time to try some, remove a jar and let it warm a bit for the oil to liquefy. If mold forms on top, remove it, and top with more oil.

Maintenance

Goats are low maintenance and pretty much take care of themselves. In fact, aside from food and shelter, human intervention boils down to caring for horns, hooves, and general health.

Dreaded Disbudding

One of the earliest and most unpleasant chores you'll face is removing the buds that will grow into horns if not attended to. The process is called disbudding, and it is no fun for the owner or the kid. However, allowing horns to grow unchecked poses a danger to yourself, other animals on your property, and the goat itself. Not only can horns lead to inadvertent goring, they can catch in fencing, trapping the goat and making it vulnerable to predators. For disbudding, a heated iron made for the purpose is held on each bud for about seven seconds. The bud is then snipped off with hoof-trimming shears. The process is quickly done, and kids seem to come through it without being worse for the wear—often feeding on hay within minutes.

Goat Keeper's Schedule

Daily
- Feed morning and evening
- Milk 12 hours apart (once a day for pygmies and dwarfs)

Weekly
- Clean pen

Monthly
- Check hooves
- Trim hooves
  6–12 months
- Deworm
  2–3 months

Diseases and Parasites

Goats are famously healthy, and while they can potentially contract many diseases, with good care and adequate feed they do remarkably well. "In 20 years of keeping goats, I've never had a disease problem," says Chuck Bauer, blaming himself for the two times he had to call in a vet—once for a gash caused by a nail on a fence post, the other when a goat got into some corn and overate. Most goat keepers will say the same. Due to the natural hardiness and fastidious eating habits of goats, they are very healthy animals. However, in addition to Enterotoxemia (See page 160.), these are the few health concerns to keep an eye on:

Worms: The need to worm varies according to pasture quality and climate. Damp pasture and humid conditions tend to breed worms goats pick up as they graze. A fecal test is the best way to determine a worm problem. Severely affected goats will lose weight, have dull hair, and generally look out of condition. Worming twice a year may be adequate in most situations. Some growers prefer herbal wormers.

Tetanus: Some growers have their flock vaccinated annually. Males are often vaccinated upon castration.

When starting out with goats, it's well worth having the vet out for a general checkup and advice on things to watch for. Surveying the many sites posted by extension services and Ag schools can easily overwhelm you with problems you'll likely never have to face. Be aware that as you gain confidence, you may well take over care and treatment yourself, as most farmers do. Know your limits, however, and call in a pro when in doubt.

Hoof Trimming

Goat hooves grow quickly, the better to cope with the wear and tear of rocky habitat. On softer soil the hooves don't get worn off and have to be trimmed every couple of months. Check hooves at least once a month, looking for growth along the edges that turns inward, sometimes to the point that the toes will begin to cross. Overgrown hooves can harbor disease, and if they really get out of hand, will cripple a goat. Trimming hooves often is better for the goat and easier on the owner. Goats don't always appreciate the favor you are doing them—finding the best way to hold the foot while working is half the battle. Trimming is best learned by having an experienced practitioner demonstrate for you, then trying it yourself.

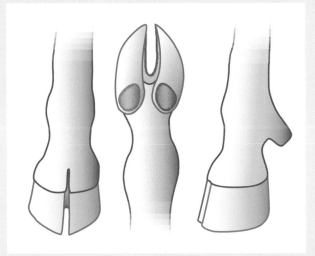

A well-trimmed hoof has no edges curling inward, no extended toes crossing, and all crevices nicely cleaned out. Check often and do a little trimming at a time.

You can buy specialty hoof cutters, but pruning shears and a pair of end-cutting pliers will do the job.

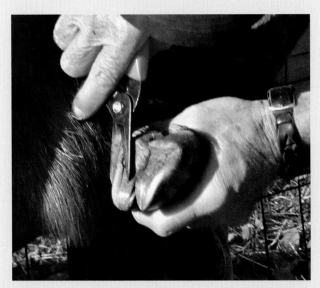

Use shears to cut off any edges that curl inward. Nip off the tips of hooves if they are overgrown and tending to cross.

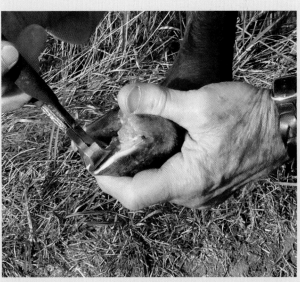

End-cutting pliers work well for nipping back the inner edge of the hoof where it is difficult to reach.

Other Livestock

Chickens and goats are often the only type of livestock allowed within municipalities. However, if you live in the country and have a couple of acres or more, here are some other livestock options to consider.

Miniature Cattle

Can you believe this? A cow not much bigger than a St. Bernard? Miniature cattle breeds are a third to half the size of full-size cattle, weighing 600–700 pounds, and produce one to four gallons of milk a day, depending on the breed. (A full-grown Holstein serves up about 10 gallons a day, a lot of milk for even a large family.) The meat is wonderful too, if you can stand to butcher an animal this cute.

Most major breeds have their miniature variants, each beautifully proportioned. Children often find them less overwhelming than their full-size cousins. Plan on at least one-half acre per animal.

These miniature Jerseys, above, live up to their breed's reputation for producing lots of milk. Despite their height of 36 to 40 in., they produce as much as 4 gal. of milk a day. Originated in Ireland, Dexter miniatures, below, weigh about 600 lbs.

Sheep

Wool, grass control, meat, and milk are what sheep have to offer. Yes, milk! Using a milking stand much like that used for a goat, you can milk a pint or two a day from a producing ewe. Depending on the fecundity of your land, up to six ewes and their lambs can be kept on one acre.

Icelandic sheep are among the hardier breeds, a helpful trait if you're getting started with sheep. They produce hardy lambs, often twins, and offer a better-than-average supply of milk. They pasture well and don't overgraze—a trait learned on Iceland's sparse vegetation. With good pasture, feed supplement is only necessary when ewes are pregnant.

Sheep need rudimentary shelter and, like goats, need their hooves trimmed regularly. To confine them, many farmers use portable electrical fencing that can be moved about the pasture area.

Pigs

Every three months, three weeks, and three days, a sow can produce a litter of 8–10 pigs. No wonder hogs have been a farming mainstay from time out of mind. Pigs are omnivorous and love kitchen slops, though it's wisest to leave meat scraps out of their diets. They also love garden weeds and grass; in fact, pasturing is one of the best and cheapest ways to bring a hog up to the optimum 200 pounds. They are masters of rooting under fences, so budget for a combination of post-and-rail fencing with welded-wire hog fence or electric fencing.

Pigs are also intelligent and, when young, about as cute as they come. You'll likely be quite attached to your pig. However, no one wants a 500-pound pet, which is what you'll have if you keep a pig too long. Even if you're resolved that slaughter is necessary, butchering a pig may not be something you want to tackle. One option is to find some local who dresses deer carcasses for hunters and hire him.

Sheep are grazing animals that can produce wool, milk, and meat simply by pasturing. Icelandic sheep (shown) are particularly suited for small farms.

With good fencing, pigs can pasture on marginal land and do a great job of rooting up and manuring gardens at season's end.

6

Beekeeping

A HIVE OR TWO OF BEES is a natural complement to a backyard homestead. Aside from the sweet deal it offers in the way of one of humankind's favorite condiments, a hive of bees guarantees pollination of your fruits, vegetables, and flowers. In addition, bee's wax makes the best possible candles, and bee venom is believed by some to have curative powers for rheumatoid arthritis, gout, and carpal tunnel syndrome.

And then there is the all-important benefit of increasing the planet's bee population, an asset currently under threat from a scourge known as Colony Collapse Disorder (CCD). This syndrome has hit commercial bees that pollinate vegetable farms especially hard. It kills whole hives of bees and threatens food production. Keeping your own hive preserves an essential resource and, as you learn the beekeeping ropes, will produce new populations. In the process, your bees will actively pollinate plants within a half-mile radius or more—a very neighborly thing to do.

And then there is the pure fascination of bees. Talk to any beekeeper long enough, and you'll catch that sense of wonder about bees. "Bees are magic," says Ernie Schmidt, a veteran beekeeper. "They have an allure. I've heard of grown men crying when their hive dies." That allure has much to do with the fact that the hive, made of thousands of members, behaves as one organism—a single-minded "one for all" attitude that runs counter to our sense of individuality. And they work so hard, diligently gathering nectar and pollen to brood their young and put something away for the winter.

Are Bees Right for You?

For the potential beekeeper, there are two deal breakers. First, if you are highly allergic to bee stings you'll likely never feel comfortable working with bees. Your doctor can test for this if you have any doubt. Second, your town or local covenants may prohibit beekeeping. You may want to appeal to your city council for an ordinance change. A nationwide trend toward tolerance is on your side. New York, San Francisco, Minneapolis, and Seattle already allow bees, and many other municipalities large and small are joining their ranks. Other things to consider include:

- **Are you reasonably handy? Most bee equipment can be bought ready-made, but you'll need to set up a stand for your hive. And who knows? Maybe someday you'll want to build your own hive.**
- **Do you have a quiet spot for a hive? It doesn't have to be in your backyard. Hives do famously well on flat rooftops.**
- **Are you comfortable with a mentor? Talk to any beekeeper, and you'll hear that joining a club and having a mentor are the best ways to learn the sometimes subtle ins and outs of beekeeping.**
- **Can you handle heavy lifting? A section of a standard hive loaded with honey can weigh 65 pounds or more. Beekeepers like to joke that there is no such thing as an old beekeeper without a bad back.**
- **Finally, are you fascinated by bees? Most beekeepers aren't in it for the honey alone, they just love the whole process.**

Bees flourish on urban rooftops. Megan Paska of Brooklyn, NY, cares for her bees within sight of the Empire State Building. One advantage: bees benefit from the residual heat of the building in winter.

How Bees Make Honey

Bees want two things. First, they want to store away enough energy in the form of honey to make it through the winter. Second, they want to raise enough new bees to swarm off and create a new colony. Bees will do anything to achieve these ends. In the peak of summer, when nectar gathering is at its most intense, a worker bee might literally wear out its wings in an effort to prepare the hive for winter—while living only six weeks.

We exploit these urges by giving bees a secure home that, just incidentally, makes it easy for us to harvest their honey. We give them plenty of sheltered space in the hive for brooding their young. When they get the urge to colonize, we generously give them additional brooder space or a new hive.

Queen Bee

It all starts with the queen. After a single mating flight with a drone, she can live up to five years and produce a million eggs. (Along the way, workers may feel she is not up to the job any more and produce a rival queen to gather a swarm about her and set out to start anew.) Workers create wax cells into which an egg and a rich blend of nectar and pollen is placed. The eggs hatch larvae that grow to be workers and drones. As the larvae vacate the cells, workers fill them with honey, capping the cells off with wax when filled. In most hives, the brooding activity works its way down the hive; storage of honey expands upward. (An exception is the Top-Bar hive, where storage expands horizontally. See page 176.)

Beekeepers harvest the stored honey by pulling out wooden frames filled with honeycomb and honey. They then separate the honey from the wax, filter it, and with no additional processing, bottle it. They leave enough honey to see the bees through the winter, supplementing it with sugar syrup if necessary.

Honey Production

1

Gather raw material. Bees are equipped like no other creature to extract nectar and pollen from flowers. A worker might visit as many as 2,000 flowers in day.

4

Gathering, gathering. Nectar flow begins with the first spring flowers. In addition to nectar, bees gather pollen, their primary protein source. In summer, 98 percent of a hive's population is involved in gathering.

2

Lay the eggs. The queen, the large bee with the bit of green paint on her back, lays eggs in the brood chamber. Workers tend to her as she lays more than 1,000 eggs a day.

3

Brood the young. Larvae are briefly fed royal jelly, then a mixture of nectar and pollen. They emerge as fully formed bees in 21–24 days.

5

Making honey. Bees use their honey stomachs to ingest and process the nectar a number of times before filling a cell. It is then capped and allowed to ripen.

6

Harvest! As the summer progresses, the beekeeper removes frames of ripe honey. The caps will be removed before the honey is processed. (See page 184.)

6 Beekeeping

Langstroth Hive

Invented in 1851 by Lorenzo Langstroth, a cleric living in Pennsylvania, the Langstroth hive is a standard hive used by beekeepers throughout much of the world. Langstroth took up beekeeping to fight depression, but he discovered that conventional practice required hacking apart the frames to get the honey out, disruptive for the bees and tough work for the beekeeper. He put his mind to improving the hive. "I have endeavored to remedy the many difficulties with which bee-culture is beset by adapting my invention to the actual habits and wants of the insect," he wrote. He noticed that bees leave "bee space" between their combs, not less than ¼ inch and not more than ⅜ inch. Not only do the bees prefer this space for moving about, it makes it easy for bee keepers to remove the comb-laden frames from the hive. Langstroth also discovered that a hive could expand with stackable boxes, a way to inhibit swarming.

The advantage of the Langstroth hive for beginners is its universality. Components are easy to buy and readily interchangeable. It has been around long enough that everyone understands how to use it. On the downside, it is heavy. A filled honey chamber called a *super* might weigh up to 75 pounds. (See illustration opposite.) One option is to use lighter 8-frame instead of the usual 10-frame boxes. The Warré hive (page 178) uses smaller boxes, and the Top- Bar hive (page 176) has no movable boxes at all.

Hives should be raised off the ground, sheltered from wind, and in a partially shaded area. Locate the hives so the flyway of bees to gathering areas doesn't intersect sidewalks, swimming pools, or your neighbor's deck.

A metal or wire-mesh mouse guard bars mice looking for a sweet treat—a particular problem in winter.

The supers of a Langstroth hive can weigh as much as 75 lbs. when loaded with honey, calling for a strong back and a helping hand.

Langstroth Hive: How It Works

The Langstroth hive comes to life when bees are introduced to the brood chamber. They soon get to work on frames filled with manufactured combs called *foundation*, helping the queen lay eggs and gathering nectar and pollen. As the brood chamber fills, workers travel up into the box above called a *super* to store honey. A screen called a queen excluder keeps the super from becoming a brooding area. As the hive fills, boxes are added above and below.

Once a super fills with honey, each frame is removed. The whole frame fits into a centrifugal extractor that spins the honey out of the comb. Fresh frames with new foundation go back into the hive to be filled with more honey.

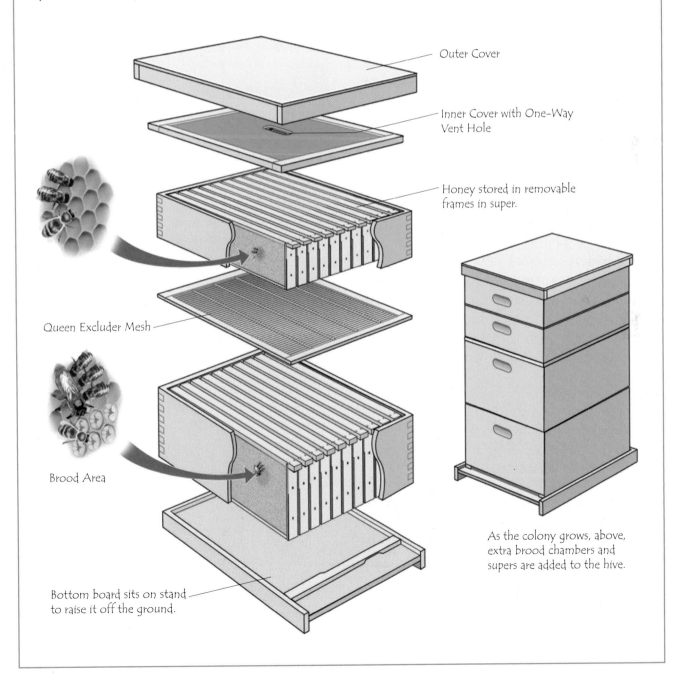

Outer Cover

Inner Cover with One-Way Vent Hole

Honey stored in removable frames in super.

Queen Excluder Mesh

Brood Area

Bottom board sits on stand to raise it off the ground.

As the colony grows, above, extra brood chambers and supers are added to the hive.

6 Beekeeping

Top-Bar Hive

The simplest hive of all to work with is the "Top-Bar" hive, also known as the Kenyan or Tanzanian hive. It is similar to the Warré hive (page 178) in its use of plain bars, but it is horizontal like a trough. Instead of the heavy lifting of removing and changing out boxes, all that is required for harvest is lifting the lidlike roof and removing a comb-laden bar. A top-bar hive can be any length or width, but one dimension is critical: the thickness of the bar. It should be 1⅜ inches wide so bees have the gap—that magic "bee space"—they prefer between the combs.

Top-bar hives produce less honey but are simple and inexpensive to build and easy to work. They are growing in popularity, but if you are a beginner you may find it difficult to find a mentor familiar with top-bar hives. Some top-bar hives locate the entry on one end of the hive; others have holes in the center of the hive. Either way, bees will build their brood nest nearest the entrance.

At harvest time, a bar with a full comb loaded with honey can be lifted out of the hive with minimal disruption to the brood area. By gently brushing away the bees, you can check for capped cells filled with ripe honey.

Top-bar hives are not difficult to build, but you may prefer to buy one ready-made. This one, the Garden Hive (**www.thegardenhive.com),** comes with a side window that lets you check progress without disturbing your bees. The basic model without a window or legs sells for less than $200; the stand with legs costs about $60. A complete starter kit with a window is about $350.

Top-Bar Hive: How It Works

Bees establish their brood nest near the hive entrance, hanging their comb from one bar. (Some beekeepers attach a strip of foundation to the bars to help things get started.) As the brood area gets established, bees store honey, working outward from the brood area.

Harvesting the honey is a simple matter of lifting out a bar and its attached honeycomb. Care must be taken to rotate the comb as you would a steering wheel (instead of end over end) to avoid breaking off the comb. Usually a honey-filled bar is left in place closest to the brooding area.

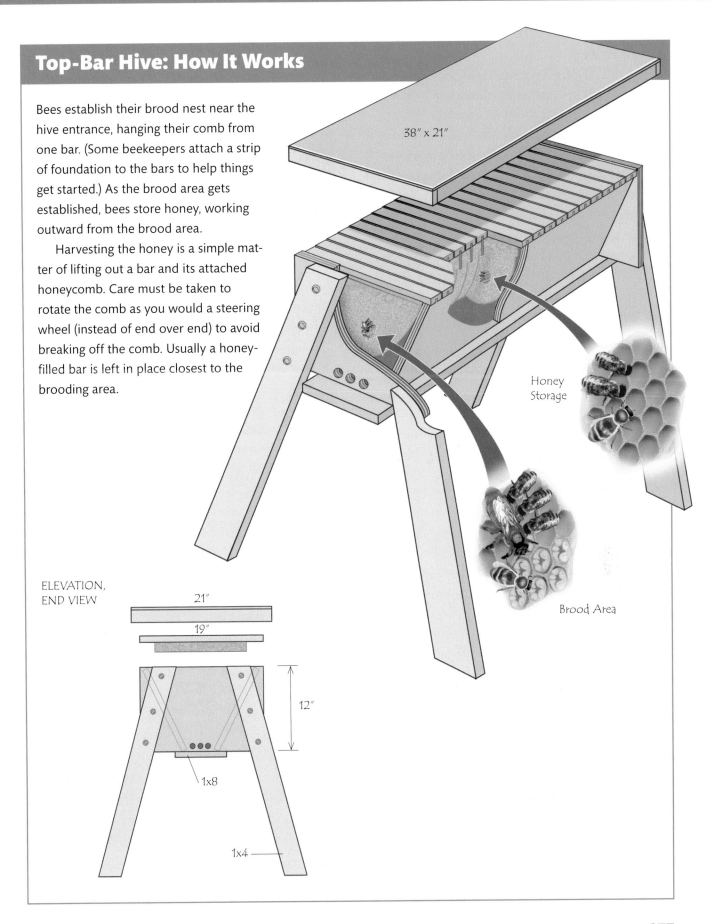

38" x 21"

Honey Storage

Brood Area

ELEVATION, END VIEW

21"

19"

12"

1x8

1x4

Warré Hive

Frenchman Abbé Emile Warré (1867–1951) experimented with more than 300 hives before he hit upon his design, which he modestly called the *ruche populaire,* or the people's hive, but which most of the world knows as the Warré hive. Warré studied hives in their natural state, especially those in sections of hollowed-out trees. In imitation of the tree, he made his hive thin and tall. He eliminated the use of manufactured foundation in rectangular frames, using instead simple bars placed a "bee space" apart from which bees built their own comb.

The Warré hive grows as the colony grows. As a brood chamber fills, another box is added underneath. If more honey storage space is needed, an empty box with bars can be slipped in place.

"The people's hive does not turn stones into honey," he wrote in his book *Beekeeping for All.* "It will not give you honey without some work. No. But the people's hive saves you a lot of expense, a lot of time, and several kilograms of honey each winter. In a word, the people's hive is a practical and logical hive. It will bring happiness to you and your bees."

Inside a hollow tree, a favorite natural home for bees, combs hang with a relatively consistent "bee space" between them. Abbé Warré set out to replicate this in his people's hive.

After removing the roof section and the quilt, it is simple to take a peek at the bars to check honey production.

Like the Top-Bar hive, the Warré hive has bars from which bees build combs. Bees are gently brushed off as the combs are harvested.

Warré Hive: How It Works

With the Warré hive, the bees are introduced to the top box where they build combs to hang beardlike from the eight bars. The sawdust-filled "quilt" box above them absorbs moisture and amplifies the scent of the bees to broadcast that this is home, a way of discouraging swarming. As new bees hatch and grow, the brooding colony moves downward. Worker bees fill the empty comb with honey. As boxes fill, new ones are added underneath, always encouraging the bees with plenty of space for expansion. In this way, honey is always rising upward, easily accessible to the beekeeper. Come winter, enough honey is left for bees to feed on as they cluster in the center of the hive to keep warm and build their resources for the coming spring.

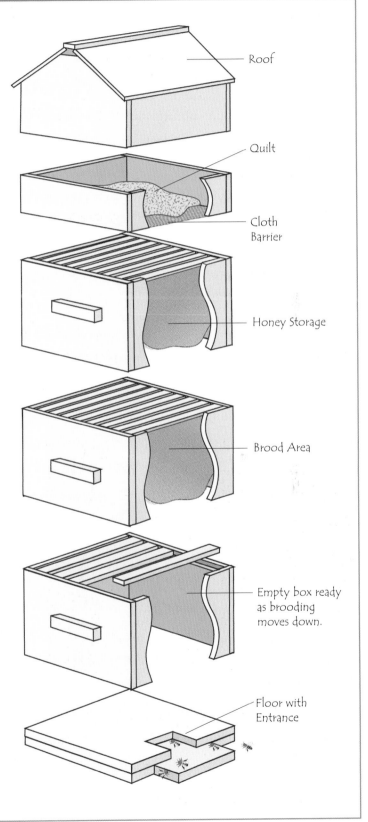

Roof

Quilt

Cloth Barrier

Honey Storage

Brood Area

Empty box ready as brooding moves down.

Floor with Entrance

A well-made Warré hive is almost a garden feature, fitting right into your backyard landscape.

Make Your Own Warré Hive

The people's hive was so named not only because it was easy to use but because it is also easy to make yourself. Begin with a web search, and select one of the several free plans available. (See **http://the-beespace.files.wordpress.com/2008/12/warre_hive_plans_english.pdf** for one good option.) What follows is an overview of the steps involved.

The adaptation shown is made of 2x10s. The thickness of the wood also makes it easier to fasten the pieces without splitting the wood or having a screw pop out the side. Cut a 1½ x ¾-inch rabbet on the ends of two of the box sides, with a ½-inch-wide x ¼-inch-deep rabbet along its top edge. Unassembled dimensions are: end-rabbeted walls, 14¹³⁄₁₆ inches long; unrabbeted walls, 13⁵⁄₁₆ inches long; all Warré boxes are 8¼ inches deep, so you will need to rip the 2x10.

Each bar also takes some prep. The bars are 12¾ inches long and 1⅜ inches wide. Cut a ½-inch-wide x ¼-inch-deep rabbet in each end. Cut a ¼-inch-deep groove down the center of each bar; glue a piece of foundation to the groove to get the bees started.

1

Cut the pieces to the needed length and width, in this case 14¹³⁄₁₆ x 8¼ in. and 13⁵⁄₁₆ x 8¼ in. Cut the necessary rabbets in the ends and along the top of the longer pieces. Predrill and start the 2-in. drywall screws.

4

Build the quilt of 1-by for a final overall dimension of 14¹³⁄₁₆ x 14¹³⁄₁₆ x 3½ in. Add screen to the bottom of the box and sawdust to form the quilt.

5

Drill a ⅛-in. hole in the ends of each bar so that it sets over the nails. Using hot wax, glue a strip of foundation into each bar. Set the bars in place.

2

Assemble the box, gluing before fastening. Complete one side; then flip the box to fasten the other side. Sand off any rough edges.

3

Mark for the nails over which the top bars will fit. Make a guide for positioning a nail every 1 in. Pound in each 1¼-in. finishing nail until the head is just below the top edge of the box.

6

Construct the bottom board. One handy feature is including a sliding tray as shown—a good way to check for mite infestation.

7

Build a final, top, box with gable ends (slope not critical), a plywood roof, and a thin ridge cap. Complete the top, and paint the hive.

Establishing a Hive

Choose your hive type, and make or buy your hive so that it is ready to go in March or so. Complete starter kits are available for Langstroth hives, including a veil, suit, gloves, hive, smoker, and feeder for under $200. Once your hive is set up, you can introduce a "package" of bees, including a queen and workers. The package, a three-pound box of about 10,000 bees, can be shipped to you, though many beekeepers prefer to pick them up directly. Once you have the bees, you can pour them into the hive. The queen is introduced more gently. She arrives in a miniature cage capped by a candy plug. Place the cage in the area where you want brooding to take place. The bees then eat through the plug to release the queen. Give the bees about four days to get established, and then check for eggs. Keep the syrup feeder well stocked until plenty of flower nectar is available.

Starting Them Well

The first year is spent establishing a healthy hive of bees. Don't expect to harvest any honey. Your main job is to keep the hive well fed and healthy—and keep track of when the bees need more space. If you see that about seven

CCD

Colony Collapse Disorder (CCD) is the name given to the sudden and mysterious death of a large number of adult bees in a colony. The exact causes are unclear. Many suspect a combination of the toxic buildup of pesticides bees are exposed to in plants, a pathogen that affects bee's digestive ability, and the stress placed upon colonies as they are trucked to pollination areas.

Backyard colonies are seldom affected by the disorder and remain of key importance in maintaining a stock of bees essential to our food supply.

of the ten frames are filled (or five of an eight-frame box), it is time to add another brood box beneath the filled one.

As the first year draws to a close and nectar gathering stops, feed the bees more sugar syrup to set them up for the winter. In cold climates, beekeepers wrap their hives with insulating blankets, being careful to allow for some ventilation. (See page 188.) In the spring of the second year, bees need feed to strengthen for the gathering season.

To start new hives, pounds of worker bees and a queen arrive in the mail in individual containers. Check that they are humming and alive. If they are banging against the screen, spritz the container with a light spray of water. The queen is likely thirsty.

Introduce the bees to their hive at dusk. Place the queen's container between frames before pouring the new bees into the hive. The container is cleverly constructed with a candy plug. Worker bees eat through it, congregating around the queen as they work.

Locating a Hive

A spot with partial shade, out of the wind, and away from damp areas is ideal for bees. Pay attention to flyways to nectar gathering areas. Make sure such pathways do not cross a deck, a neighbor's hot tub, or a nearby play-ground. Interrupting a flyway with a high fence puts bees at an altitude that takes them over people's heads. Some beekeepers camouflage their hive so as not to cause undue alarm among the neighbors.

Hives are painted mainly to preserve the wood. (The interior is unpainted.) Color is open to debate. One school of thought says camouflage them with dark colors to calm neighbors. Another school asserts that bold, colorful patterns help bees find their hives. Or maybe the blossom tones above are best? The choice is up to the beekeeper.

Harvesting Honey

Toward the end of the summer you can check the supers for their honey supply. Look for capped cells of honey—uncapped nectar-filled cells are not ripe for harvesting. Some beekeepers use a noxious fluid to flush the bees from the super; others find that placing the inner cover with its one-way vent (see page 175) lets the bees exit the super but not get back in. Once the frames are removed, uncap the cells with a flat, hot electric knife designed for the purpose. Once both sides of the frame are uncapped you can place the frames in the separator.

One of the benefits of a beekeeping club is the free use of a separator and other honey-processing equipment—a big advantage because even a small separator costs about $300 and is only used one season of the year. The separator spins the frames, forcing the honey out of the comb. Running the honey through a mesh as it is emptied from the spigot into a bucket removes any bits of wax. Jars can be filled from a spigot at the bottom of the bucket. (Beekeepers love spigots—much less messy than pouring into funnels.)

1

Honey harvest begins by lifting off the cover and quieting the bees with some smoke. The frames are laden with honey and bee's wax (inset). The beekeeper carefully cuts and pries them apart so that single frames can be removed.

4

Once the frames are set in the separator, the lid is closed and the frames are rotated at high speed. The honey releases from the comb by centrifugal force and drips down the walls of the separator.

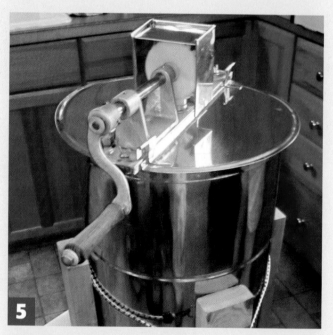

5

This separator is owned by a bee club, a great way to save on buying an expensive piece of equipment that might be used only twice a year.

2

Next, the beekeeper removes each frame, brushing off the bees on it. Ripe honey is uniformly capped with wax, filling the interior of the frame.

3

A heated blade pushed across the surface of the comb neatly peels off the caps (inset). These uncapped frames are loaded with honey and ready for the separator. Remove caps on both sides of the comb.

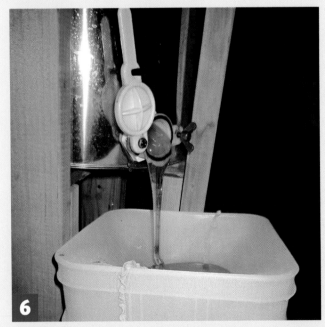

6

Mesh over the top of the collector bucket catches any wax and debris. This bucket has a spigot, a much neater way of filling jars than using a funnel.

7

Almost any clean glass jar can be used for honey. Some beekeepers prefer canning jars for storage because they are large and uniform.

Organic Honey

Hankering to produce certified organic honey? It can be tough. According to standards set by Quality Assurance International, an independent, third-party certifier of organic products, the foraging requirements are daunting. Because bees range so far in their nectar gathering, to be certified organic your honey must be produced by naturally foraging colonies "that are located at least 2 miles distant, in straight-line flight, from any pollution source which could cause the honey to become contaminated by, or as a result of, returning foraging bees (e.g. synthetic-chemical sprayed agriculture, industrial centers, urban centers, etc.)." If your backyard is in the middle of the wilderness, you're in good shape. Otherwise, you'll be hard put to meet the standards.

Because your bees will forage far and wide, it is difficult to meet the requirements for organic honey.

Tips for Cooking with Honey

- Lower baking temperature by 25°. Honey makes baked goods brown faster.
- Measure honey easily by coating cups or spoons with oil or nonstick spray.
- Substitute ⅔ to ¾ cup of honey per cup of sugar (depending on taste).
- Decrease the amount of liquids by ¼ cup per cup of honey used.
- Store honey at room temperature.
- If honey crystallizes, remove lid and place jar in warm water until crystals dissolve, or microwave honey on HIGH for 2 to 3 minutes or until crystals dissolve, stirring every 30 seconds. Do not scorch.
- Honey should not be fed to babies less than one year of age. Honey is a safe and wholesome food for older children and adults.

Courtesy of the National Honey Board, **www.honey.com**

Other Stuff from Bees

Pollen: Each hive collects more than 60 pounds of pollen per year. Pollen is larvae food, containing up to 35 percent protein and 10 percent sugars, as well as carbohydrates, enzymes, and several important vitamins and minerals.

Propolis: A sticky resin collected from tree sap and mixed with wax, it is the glue that holds the hive together. Bees use it to seal cracks and make repairs. Humans use it as the base for fine varnish.

Royal Jelly: Made of digested pollen and honey or nectar mixed with bee-secreted chemicals, royal jelly is the stuff that queens are made of, though all bees feed on it briefly. Many people find cosmetic and health benefits in royal jelly.

Working with Bees

Perhaps it is inbred defense designed to fight off invading bears, but bees tend to go for your face when they're in a stinging mood. That's why some beekeepers forego a protective suit and gloves but always wear face protection. However, to be fully protected, beekeepers don full gear with care taken to keep bees from wandering up ankles or wrists. Stings are an occupational hazard. If stung, pull out the stinger to stop the flow of venom. The stings hurt, and the welt will linger for a few days, but the pain dissipates pretty quickly.

A full beekeeper's veil and suit is wise if you are new to beekeeping or when bees get particularly riled up. At other times, only a veil is necessary.

The Beekeeper's Friend

The smoker is the beekeeper's friend. This simple device is basically a canister with a bellows attached. Stocked with any sort of slow burning material—wood chips, burlap, bark—it creates a smokescreen that works on bees in two ways. First, it seems to send them into forest-fire response. They gorge themselves with honey and get stupefied in the process. Second, the smoke disturbs the chemical smells by which bees communicate. Confusion reigns, and bees are unable to make a concerted attack. The smoker helps whenever you open the hive, though some beekeepers have developed such a calm, quiet technique that they seldom need it.

Winterizing a Hive

Bees cluster together for warmth in the winter, feeding on honey and taking a well-deserved rest before reigniting brooding in the spring. Beekeepers leave one or two supers filled with honey for the use of the bees, sometimes supplementing the reserves with syrup. Some beekeepers wrap hives with tar paper or slip on an insulated cover manufactured for the purpose. In cold climates, the hives are encased in a close-fitting plywood box, called a false backing, wrapped with insulation and plastic sheets. In all cases, a small hole must be left at the top of the hive to vent moisture. At the bottom of the hive, cover the entrance with hardware cloth or a perforated metal mouse guard. (See page 174.)

The degree of winterizing varies according to climate. In moderate climes, little or no wrapping is necessary. These hives have to go through a cold, snowy winter and are dressed accordingly. Note the gaps in the coverage to allow ventilation.

Diseases and Parasites

Like any organism, a colony of bees can fall prey to disease and parasites. Just paying attention is one of the best preventive measures, observing that bees are active at the hive entry, that brood and honey cells look healthy and nicely fleshed out, and that there are no odd smells. Keeping bees fed year-round and guarding the hive from dampness also help maintain health. One of the more prevalent diseases is American Foul Brood, which attacks larvae. Preventive measures include never feeding bees honey, avoiding materials that might be infected, and regular hive maintenance so that signs of the disease can be detected early. To defend larvae from the disease, many beekeepers administer Terramycin in powder or patty form in the spring and fall. Two types of mites can infest hives. The main defenses against mites include treatments of menthol, various chemicals, and techniques that interrupt the life cycle of the parasites. There are other diseases and parasites that can bother bees, but before you scare yourself by what your bees might contract, get advice from local beekeepers on what to guard against in your region.

Capturing a Swarm

Many beekeepers advertise that they will collect nuisance swarms—a great way to get bees for free. However, few people discriminate between honeybees, bumblebees, yellow jackets, and hornets, so many beekeepers advertise that they will remove honeybees for free but will charge for all other types. A bee swarm will attach to a tree limb (or a picnic bench or a tractor tire—almost anything really) in a dense cluster. Often it is fairly close to the ground. Swarms are quiet in the morning or evening. Swarm-capture technique involves suiting up, placing a hive chamber or cardboard box directly beneath the swarm, and with a sharp shake, dropping the swarm into the box. Gently pour all the bees you can into a brood box, and gently set the frames in place. If the cluster attaches to a tree or bush, you can sometimes cut away the cluster and set it, branches and all, in a box. Put a lid on the box, and leave it overnight to attract stragglers before wrapping up the box and hauling it home. Help an old hand do this before trying it yourself.

This swarm was discovered as a pendulous wad of bees hanging in a tree just a few feet off the ground. The bees are clustered around the queen and are docile.

An experienced beekeeper volunteered to remove the swarm. First step: place a hive chamber as close to the swarm as possible.

With protective head gear in place, it's time for a quick shake of the branch to dump the bees onto the hive.

Hopefully, the chamber is just what the bees have been looking for, and they'll gather in the hive. Some gentle brushing encourages stragglers.

7 Harvest Home

RAISING EVEN A SMALL AMOUNT of your own food is an important step toward self sufficiency, but for your backyard homestead to keep delivering food into the winter, you'll need to master some preserving techniques. This chapter introduces you to the joys of canning, drying, pickling, smoking, and brewing. In addition, you'll learn the all-important art of storing your preserved harvest safely and securely. It all adds up to keeping your harvest in the healthiest and most cost-efficient way.

Canning

Summer in a jar. That's what you'll experience when you pop open the lid of your lovingly canned produce. Canning is one of the most popular ways to preserve your garden bounty. With a modest investment in equipment, you can safely put up your produce for the winter, and in the process save yourself up to half the cost of commercially canned food. And almost anything can be canned, not just vegetables and fruit, including meat and fish—in fact, the tenderness and flavor of canned beef is something fondly remembered by farm-raised folk.

Why Canning Is Necessary

Food is perishable because its water content fosters the growth of undesirable microorganisms such as bacteria, molds, and yeasts. It also spoils because of the activity of enzymes and reactions with oxygen.

Canning, properly done, not only preserves food but also guards against that most fearsome of contaminants, botulism. Botulism is a deadly form of food poisoning resulting from the growth of the bacterium *Clostridium botulinum*. This bacterium survives harmlessly in soil and water, but when conditions are right, it multiplies rapidly, producing a deadly toxin as it does. Contact with botulinum toxin can be fatal whether it is ingested or enters through the skin. Effective canning practices include

- **Carefully selecting and washing fresh food.**
- **Peeling when indicated.**
- **Hot packing. (See page 193.)**
- **Adding acids (lemon juice or vinegar).**
- **Using the proper jars and self-sealing lids to create a vacuum in the jars.**
- **Processing jars in a boiling-water or pressure canner for a specified period of time.**

Canning Methods

Boiling-water canning involves placing food in sealed jars and immersing the jars in boiling water for a specified amount of time. **Pressure canning** is a similar process but uses pressure to boost the heat above boiling temperature to kill microorganisms. The method used depends on the acidity of the food.

Boiling-water canning produces enough heat to kill botulinum spores in high-acid foods. Low-acid foods require a temperature well above the boiling point, 240°F to 250°F, attainable only with pressure canners. The exact time for both methods depends on the kind of food being canned, the way it is packed into jars, and the size of the jars.

Fruit often seems to ripen too quickly. Jellies and preserves put it to good use so that summer's bounty can be enjoyed through the winter.

Two Canning Methods

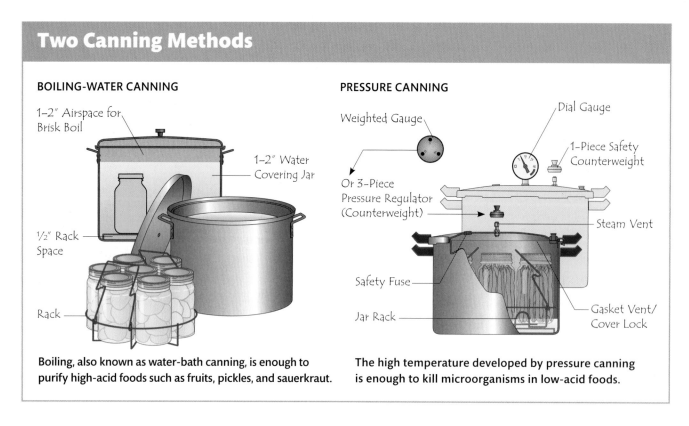

BOILING-WATER CANNING

1–2″ Airspace for Brisk Boil

1–2″ Water Covering Jar

½″ Rack Space

Rack

PRESSURE CANNING

Weighted Gauge

Dial Gauge

Or 3-Piece Pressure Regulator (Counterweight)

1-Piece Safety Counterweight

Steam Vent

Safety Fuse

Jar Rack

Gasket Vent/ Cover Lock

Boiling, also known as water-bath canning, is enough to purify high-acid foods such as fruits, pickles, and sauerkraut.

The high temperature developed by pressure canning is enough to kill microorganisms in low-acid foods.

The Importance of Being Acid

Acidity is the continental divide of canning. It determines what method (boiling-water or pressure canning) is appropriate to control botulinum bacteria in a given food. Acidity is measured by its pH; the lower its pH value, the more acid the food. Acid foods have a pH of 4.6 or lower. Acid foods contain enough acid to block the growth of botulinum bacteria or to destroy them more rapidly when heated in hot-water canning. Low-acid foods like meats and all fresh vegetables are not acidic enough to prevent the growth of these bacteria.

Altitude Matters

Using the process time for canning food at sea level may result in spoilage if you live at altitudes of 1,000 feet or more. Water boils at lower temperatures as altitude increases. Lower boiling temperatures are less effective for killing bacteria. At higher elevations, boil one additional minute for each additional 1,000 feet of elevation.

Ensuring High-Quality Canned Foods

Begin with good-quality fresh foods suitable for canning. Discard diseased and moldy food. Process most fruits and vegetables within 6 to 12 hours after harvest. For best quality, apricots, nectarines, peaches, pears, and plums should be ripened one or more days between harvest and canning. If you must delay the canning of other fresh produce, keep it in a shady, cool place. Follow these guidelines:

- **Use only foods that are at the proper maturity and are free of diseases and bruises.**
- **Use the hot-pack method, especially with acid foods to be processed in boiling water.**
- **Don't expose prepared foods to air unnecessarily. Can them as soon as possible.**
- **While preparing jars, keep peeled, halved, quartered, sliced, or diced apples, apricots, nectarines, peaches, and pears in a solution of 1 teaspoon (3g) ascorbic acid to 1 gallon of cold water. This also maintains the natural color of mushrooms and potatoes and prevents stem-end discoloration in cherries and grapes.**

Packing Methods

Raw packing is the practice of filling jars tightly with freshly prepared, unheated food. Entrapped air in and around the food may cause discoloration within two to three months of storage. Raw packing is more suitable for vegetables processed in a pressure canner.

Hot packing involves boiling freshly prepared food, simmering it two to five minutes, and promptly filling jars loosely with the boiled food. Hot packing is the best way to remove air and is the preferred packing style for foods processed in a boiling-water canner.

With both methods, any added juice, syrup, or water should be boiled first. Added liquid helps to remove air from food tissue, shrinks food, helps keep the food from floating in the jars, increases vacuum in sealed jars, and improves shelf life.

With the raw-pack method, boiling water is added to raw vegetables. A pressure canner is the safest canner for this approach.

Hot packing uses precooked produce.

Sources of Ascorbic Acid

Ascorbic acid helps preserve natural colors in food. You can get ascorbic acid in several forms:

- *Powdered* ascorbic acid is seasonally available among canners' supplies in supermarkets or online. One level teaspoon of pure powder weighs about 3 grams. Use 1 teaspoon per gallon of water as a treatment solution.

- *Vitamin C tablets* are an economical source available year-round. Buy 500-milligram tablets. Crush and dissolve six tablets per gallon of water.

- *Commercially prepared mixes* of ascorbic and citric acid are available seasonally. Sometimes citric acid powder is sold in supermarkets, but it is less effective in controlling discoloration. If you choose to use these products, follow the manufacturer's directions.

Use Current Recipes

Always use current published instruction and recipe manuals. Family recipes that have been handed down through the years may hold sentimental value, but they may be unreliable and usually do not include tested processing pressures and times vital to successful and safe canning. Extensive research has been conducted on canning in recent years. Canning information published prior to 1988 may be incorrect and could pose a serious health risk. In addition, every fruit or vegetable requires a specific amount of time in the canner for safe results. Consult the series of guides published by the USDA and available online through the National Center for Home Food Preservation at **www.uga.edu/nchfp/publications/publications_usda.html.** You will learn how much produce is needed per canner load, what to look for when harvesting, how to prep the food, and how to adjust the canner pressure to compensate for your altitude.

Jars and Lids

Use only Mason-type, threaded, home-canning jars made of heat-tempered glass. Do not use recycled pickle and mayonnaise jars. Canning jars are available in ½-pint, 1-pint, 1½-pint, 1-quart, and ½-gallon sizes. The standard jar-mouth opening is about 2⅜ inches. Wide-mouth jars have openings of about 3 inches. Half-gallon jars may be used for canning very acid juices. Regular-mouth decorator jelly jars are available in 8- and 12-ounce sizes.

Washing Jars

Before every use, wash empty jars in hot water with detergent by hand or in a dishwasher. Don't use abrasive pads to clean the jars. Rinse well; detergent residues may cause unnatural flavors and colors. Keep jars hot until ready to fill with food. Submerge the clean empty jars in enough water to cover them in a large stockpot or boiling water canner. Bring the water to a simmer (180°F), and keep the jars in the simmering water until it is time to fill them with food. You may use a dishwasher for preheating jars if they are washed and dried in a complete regular cycle. Keep the jars in the closed dishwasher until needed for filling.

You can wash home canning jars and use them repeatedly as long as the rim is free of nicks. Remove screw bands after canning and reuse. (The vacuum seal holds the lid on.) Buy lids fresh each season and never reuse them.

Sterilizing Empty Jars

Sterilize empty jars used for all jams, jellies, and pickled products processed less than 10 minutes. To sterilize empty jars after washing, submerge them, right side up, in a boiling-water canner with the rack in the bottom. Fill the canner with enough warm water so that it is 1 inch above the tops of the jars. Bring the water to a boil, and boil 10 minutes. Reduce the heat under the canner, and keep the jars in the hot water until it is time to fill them. Remove and drain hot steril-ized jars one at a time, saving the hot water in the canner for processing filled jars. Fill the sterilized jars with food; add lids; and tighten screw bands.

Empty jars used for vegetables, meats, and fruits to be processed in a pressure canner do not feed to be sterilized. It is also unnecessary to sterilize jars for fruits, tomatoes, and pickled or fermented foods that will be processed 10 minutes or longer in a boiling-water canner.

All About Canning Lids

When jars are processed, the lid gasket softens and flows slightly to cover the jar-sealing surface yet allows air to escape from the jar. The gasket forms an airtight seal as the jar cools. Gaskets in unused lids work well for at least 5 years from date of manufacture, but try not to buy more lids than needed for a year. Don't use older, dented, or deformed lids.

Put the lids in a saucepan, and cover them with water. Simmer, but do not boil. Reduce heat, and keep them hot until you need them. Release air bubbles in the jar by inserting a flat plastic spatula between the food and the jar. Clean the jar rim with a dampened paper towel. Place the preheated lid, gasket down, onto the jar, and fasten on the screw band.

Do not retighten lids after processing the jars. As jars cool, a vacuum is created that seals the lid. If rings are too loose, liquid may escape from the jars during processing, and the seals may fail. If rings are too tight, air cannot vent during processing, and food will discolor during storage. Over-tightening also can cause lids to buckle and jars to break. Remove the screw band after the jar cools. Wash and dry it for reuse.

Make It Easy on Yourself

The right tools make the job easier. Here are a few handy canning aids.

Lids must be sterilized in hot, not boiling, water. Use a magnetic lid lifter to remove them from the pan.

Air trapped between chunks of food can lead to discoloration and spoilage. The lid lifter doubles as a blade for easing food away from the inside of the jar, removing air pockets.

Spare yourself frustration and mess in the heat of battle with this canning-jar-sized funnel.

Using Boiling-Water Canners

Although pressure canners may also be used for processing acid foods, boiling-water canners are faster. A pressure canner would require from 55 to 100 minutes to process a load of jars; while the total time for processing most acid foods in boiling water is only 25 to 60 minutes.

1

Before you start preparing your food, fill the canner halfway with clean water. This is approximately the level needed for a canner load of pint jars. For other sizes and numbers of jars, the amount of water in the canner will need to be adjusted so that it will be 1 to 2 in. over the top of the filled jars.

2

Preheat water to 140°F for raw-packed foods and to 180°F for hot-packed foods. Food preparation can begin while this water is preheating.

5

Set a timer for the total minutes required for processing the food.

6

Keep the canner covered, and maintain a boil throughout the processing time the recipe stipulates. **IMPORTANT:** if the water stops boiling at any time during the process, bring the water back to a vigorous boil and *begin the timing of the process over, from the beginning.*

3

Load filled jars, fitted with lids, into the canner rack, and use the handles to lower the rack into the water. Or fill the canner with the rack in the bottom, one jar at a time. When using a jar lifter, position it below the neck of the jar (below the screw band of the lid). Keep the jar upright. Tilting the jar could cause food to spill into the sealing area of the lid.

4

Add more boiling water, if needed, so that the water level is at least 1 in. above jar tops. For process times over 30 minutes, the water level should be at least 2 in. above the tops of the jars. Turn the heat to its highest position; cover the canner with its lid (inset); and heat until the water in the canner boils vigorously.

7

Add more boiling water, if needed, to keep the water level above the jars. When jars have been boiled for the recommended time, turn off the heat, and remove the canner lid (inset). Wait five minutes before removing jars.

8

Using a jar lifter, remove the jars and place them on a towel, leaving at least 1-in. spaces between the jars during cooling. Let the jars sit undisturbed to cool at room temperature for 12 to 24 hours.

Using Pressure Canners

Low-acid foods must be processed in a pressure canner to be free of botulism risks. It alone develops a high enough temperature to sterilize low-acid food. Pressure canning can be used with acid foods as well.

1 **Carefully clean the pressure canner;** make sure the vent port is clean. A pipe cleaner will clear it (inset). Place the rack in the bottom of the pressure canner, and add boiling water, following the manufacturer's recommendations.

2 **Place filled jars** on the rack, using a jar lifter. When using a jar lifter, make sure it is securely positioned below the neck of the jar (below the screw band of the lid). Keep the jar upright at all times. Tilting the jar could cause food to spill into the sealing area of the lid. Fasten the canner lid securely.

5 **Regulate heat** to maintain a steady pressure at or slightly above the correct gauge pressure. Follow the canner manufacturer's directions. **IMPORTANT:** if at any time pressure goes below the recommended amount, bring the canner back to pressure and *begin the timing of the process again.*

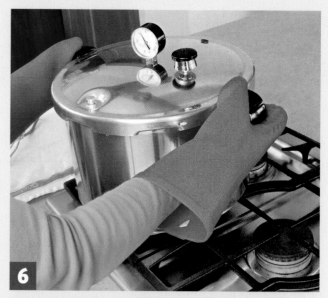

6 **When the timed process is completed,** remove the canner from heat and let the canner depressurize. Do not force-cool. Cooling with cold running water or opening the vent port may result in unsafe food. Older heavy-walled canners may take up to 45 minutes to depressurize. Newer thin-walled canners cool more rapidly.

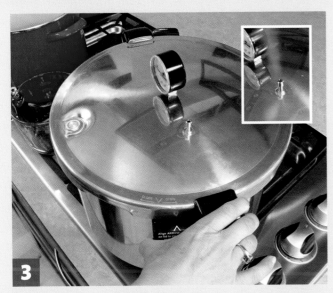

3

Carefully seat and lock the lid, following the manufacturer's requirements. Leave the weight off the vent port or open petcock. Heat at the highest setting until steam flows freely from the open petcock or vent port (inset).

4

While maintaining the high-heat setting, let the steam exhaust continuously for 10 minutes, and then place the weight (also called a pressure regulator) on the vent port or close the petcock. Start timing the process when the pressure reading on the dial gauge (inset) indicates that the recommended pressure has been reached.

7

After the canner is depressurized, remove the weight from the vent port or open the petcock. Wait 10 minutes; unfasten the lid; and remove it carefully. Lift the lid away from you so that the steam does not burn your face.

8

Remove jars with a jar lifter, and place them on a towel, leaving at least 1-in. spaces between the jars during cooling. Let the jars sit undisturbed to cool at room temperature for 12 to 24 hours.

Pressure Canners

Modern pressure canners are lightweight, thin-walled kettles; most have turn-on lids. They have a jar rack, gasket, dial or weighted gauge, an automatic vent/cover lock, a vent port (steam vent) to be closed with a counterweight or weighted gauge, and an over-pressure plug.

The manufacturer's directions will explain how to operate the appliance. Because maintaining pressure is so important in canning, pay careful attention to suggestions for reading and adjusting the pressure.

Leaving Headspace

Headspace is the air space between the top of the food or its liquid and the lid. All recipes specify the amount of headspace necessary for the food being canned. Too much headspace can result in under processing because it may take too long to release the air from the jar. Too little headspace will trap food between the jar and the lid and may result in an inadequate seal. As a general rule, allow a ½-inch headspace for fruits and tomatoes and a 1-inch space for all vegetables, meats, poultry, and seafood.

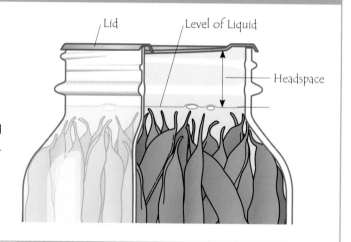

Storing Canned Foods

Label and date the jars, and store them in a clean, cool, dark, dry place. Do not store jars above 95°F or near hot pipes, a range, a furnace, under a sink, in an uninsulated attic, or in direct sunlight. Dampness may corrode metal lids, break seals, and allow recontamination and spoilage.

Accidental freezing of canned foods will not cause spoilage unless jars unseal. However, freezing and thawing may soften food. If jars must be stored where they may freeze, wrap them in newspaper; place them in heavy cartons; and cover with more newspapers and blankets.

Cool, dark, and dry are the bywords for storing canned goods. If stored in a basement, keep canned goods away from areas of fluctuating heat such as a laundry area or near a furnace.

Food Gone Bad

Never, ever sample food from a jar with an unsealed lid or food that shows signs of spoilage. Instead, every time you reach for some canned produce do the following:

1. Check the lid for swelling. Also look for a broken seal—easy to do if you've removed the screw band before storage. A bulging lid or broken seal is the result of the gas caused by spoilage. Lids with concave centers are sealed.

2. Check the outside of the jar for streaks of dried food. Look at the contents for rising air bubbles and unnatural color.

3. While opening the jar, spurting liquid or a bad odor is a warning sign. Look for cottonlike white, blue, black, or green mold growth on the food surface and underside of the lid.

4. Contaminated low-acid foods, including tomatoes, may show little sign of spoilage. Treat all suspect containers as if they have botulinum toxin. Deal with them as follows:

■ If the suspect glass jar is still sealed, place it in a heavy garbage bag and place in trash container.

■ If the suspect glass jars are unsealed, open, or leaking, they should be detoxified before disposal. Wear disposable rubber or heavy plastic gloves. Carefully place the sealed containers on their sides in an 8-quart or larger stockpot, pan, or boiling-water canner. Wash your gloved hands thoroughly. Carefully cover the containers with at least 1 inch of water, and avoid splashing the water. Cover the pot, and boil for 30 minutes. Cool and discard the containers, their lids, and food.

■ Wearing gloves, clean the area. (Remember: botulinum toxin can be fatal even by contact with your skin.) Use a solution of 1 part chlorine bleach to 5 parts clean water to treat anything that might have been contaminated, including clothing. Spray or wet contaminated surfaces with the bleach solution, and let it stand for 30 minutes. Wipe the solution up with paper towels, putting the towels in a plastic bag before disposing of them. Repeat the process. Discard the gloves when the cleaning process is complete.

Testing Jar Seals

After cooling jars for 12 to 24 hours, remove the screw bands and check the seals using one of these methods.

Press the middle of the lid with a finger or thumb. If the lid springs up when you release your finger, the lid is unsealed.

Tap the lid with the bottom of a teaspoon. If it makes a dull sound, the lid is not sealed. If food is in contact with the underside of the lid, it will also cause a dull sound. If the jar is sealed correctly, it will make a ringing, high-pitched sound.

Hold the jar at eye level and look across the lid. The lid should be concave (curved down slightly in the center). If the center of the lid is either flat or bulging, it may not be sealed.

Preparing Pickled and Fermented Foods

Everyone understands the concept of pickled foods, but fermented? Lacto-fermentation happens when friendly bacteria, *lactobacilli,* are introduced and consume the starches and sugars in foods, producing tangy and preservative lactic acid and adding beneficial bacteria to the human digestion system. (There is evidence that our lack of such bacteria is why lactose and gluten intolerance, digestive irritation, yeast infections, allergies, and asthma are so rampant.) The preservation process is quick. Regular dill pickles and sauerkraut are fermented and cured for about three weeks. Refrigerator dills are fermented for about one week. During curing, colors and flavors change and acidity increases.

Pickling relies on salt and vinegar for preservation. Fresh-pack or quick-process pickles are not fermented; some are brined several hours or overnight, then drained and covered with vinegar and seasonings. Fruit pickles are usually prepared by heating fruit in seasoned syrup acidified with either lemon juice or vinegar. Relishes are made from chopped fruits and vegetables that are cooked with seasonings and vinegar.

Ingredients

Select fresh, firm fruits or vegetables that are free of spoilage. Choose a recipe, and measure or weigh amounts carefully because the proportion of fresh food to other ingredients will affect flavor and, in many instances, safety.

Use canning or pickling salt. Noncaking material added to other salts may make the brine cloudy. Because flake salt varies in density, it is not recommended for making pickled and fermented foods. For sweetening, white granulated and brown sugars are most often used. Corn syrup and honey, unless called for in reliable recipes, may produce undesirable flavors. White distilled and cider vinegars of 5-percent acidity (50 grain) are recommended. White vinegar is usually preferred when light color is desirable, as is the case with fruits and cauliflower.

Pickling is a good example of a preservation method that has given birth to hundreds of delicious foods.

Pickles with Reduced Salt Content

In the making of fresh-pack pickles, cucumbers are acidified quickly with vinegar. Use only tested recipes formulated to produce the proper acidity. While you may prepare these pickles safely with reduced or no salt, their quality may be noticeably lower. Both texture and flavor may be slightly, but noticeably, different than expected. You may wish to make small quantities first to determine whether you like them or not.

However, the salt used in making fermented sauerkraut and brined pickles not only provides characteristic flavor but is vital to safety and texture. In fermented foods, salt favors the growth of desirable bacteria while inhibiting the growth of others. **Caution:** do not attempt to make sauerkraut or fermented pickles by cutting back on the salt required.

Pickling is not only prudent, but presentable. A combination of vegetables serves as a beautiful and delicious side dish.

Making a Batch of Fermented and Pickled Vegetables

A 1-gallon container is needed for each 5 pounds of fresh vegetables. Therefore, a 5-gallon stone crock is of ideal size for fermenting about 25 pounds of fresh cabbage or cucumbers. Food-grade plastic and glass containers are excellent substitutes for stone crocks. You may use other 1- to 3-gallon nonfood-grade plastic containers if you line the inside with a clean food-grade plastic bag. **Caution:** be certain that foods contact only food-grade plastics. Do not use garbage bags or trash liners. Fermenting sauerkraut in 1-quart and ½-gallon Mason jars is an acceptable practice but may result in more spoilage losses.

Cabbage and cucumbers must be kept 1 to 2 inches under the brine while fermenting. After adding prepared vegetables and brine, insert a suitably sized dinner plate or glass pie plate inside the fermentation container. The plate must be slightly smaller than the container opening yet large enough to cover most of the shredded cabbage or cucumbers. To keep the plate under the brine, weight it down with two to three sealed quart jars filled with water. Covering the container opening with a clean, heavy bath towel helps to prevent contamination from insects and mold while the vegetables are fermenting. Fine-quality fermented vegetables are also obtained when the plate is weighted down with a very large, clean plastic bag filled with 3 quarts of water containing 4½ tablespoons of canning or pickling salt in case the bag leaks. Freezer bags sold for packaging turkeys are suitable for use with 5-gallon containers.

The fermentation container, plate, and jars must be washed in hot sudsy water and rinsed well with very hot water before use.

Preparing Butters, Jams, Jellies, and Marmalades

Sweet spreads consist of fruits preserved mostly by means of sugar and a thickening or jelling agent. Fruit jelly is a semisolid mixture of fruit juice and sugar that is clear and firm enough to hold its shape. Other spreads are made from crushed or ground fruit.

Jam also will hold its shape, but it is less firm than jelly. Jam is made from crushed or chopped fruits and sugar. Jams made from a mixture of fruits are usually called conserves, especially when they include citrus fruits, nuts, raisins, or coconut. Preserves are made of small, whole fruits or uniform-size pieces of fruits in a clear, thick, slightly jellied syrup. Marmalades are soft fruit jellies with small pieces of fruit or citrus peel evenly suspended in a transparent jelly. Fruit butters are made from fruit pulp cooked with sugar until thickened to a spreadable consistency.

Ingredients

For proper texture, jellied fruit products require the correct combination of fruit, pectin, acid, and sugar. The fruit gives each spread its unique flavor and color. It also supplies the water to dissolve the rest of the necessary ingredients and furnishes some or all of the pectin and acid. Good-quality, flavorful fruits make the best jellied products.

Pectins jell if they are in the right combination with acid and sugar. All fruits contain some pectin. Apples, crab apples, gooseberries, and some plums and grapes usually contain enough natural pectin to jell. Other fruits, such as strawberries, cherries, and blueberries, contain little pectin and must be combined with other fruits high in pectin or with commercial pectin products to jell. Because fully ripened fruit has less pectin, one-fourth of the fruit used in making jellies without added pectin should be underripe.

The proper level of acidity is critical to jelly formation. If there is too little acid, the jelly will never set; if there is too much acid, the jelly will lose liquid (weep). For fruits low in acid, add lemon juice or other acid

Fresh fruit comes and goes all too quickly. Sweet spreads are a traditional way to preserve it.

ingredients as directed. Commercial pectin products contain acids that help to ensure jelling.

Sugar serves as a preserving agent, contributes flavor, and aids in jelling. Cane and beet sugar are the usual sources of sugar for jelly or jam. Corn syrup and honey may be used to replace part of the sugar in recipes, but too much will mask the fruit flavor and alter the jelly structure. Use tested recipes for replacing sugar with honey and corn syrup. Do not try to reduce the amount of sugar in traditional recipes. Too little sugar prevents jelling and may allow yeasts and molds to grow.

No Wax

Even though sugar helps preserve jellies and jams, molds can grow on the surface of these products. The mold that people usually scrape off the surface of jellies may not be as harmless as it seems. Mycotoxins have been found in some jars of jelly having surface mold growth. Mycotoxins are known to cause cancer in animals; their effects on humans are still being researched.

Because of possible mold contamination, paraffin or wax seals are no longer recommended for any sweet spread, including jellies. To prevent growth of molds and loss of good flavor or color, fill hot products into sterile Mason jars, leaving a ¼-inch headspace; seal with self-sealing lids; and process 5 minutes in a boiling-water canner. Adjust process time for higher elevations. If unsterile jars are used, the filled jars should be processed for 10 minutes, but the added 5-minute process time may cause weak jellies. (For how to sterilize empty jars, see page 195.)

Making Jams and Jellies

The standard method, which does not require added pectin, works best with fruits naturally high in pectin. The other method, which requires the use of commercial liquid or powdered pectin, is much quicker. The jelling ability of various pectins differs. To make uniformly jelled products, be sure to add the quantities of commercial pectins to specific fruits as instructed on each package. Overcooking may break down pectin and prevent proper jelling. When using either method, make one batch at a time, according to the recipe. Increasing the quantities often results in soft jellies. Stir constantly while cooking to prevent burning. Recipes are developed for specific jar sizes. If jellies are filled into larger jars, excessively soft products may result. (Check USDA Guide *Preserving Food: Jams and Jellies*, for specific recipes with processing times.)

Adding syrup to canned fruit helps to retain its flavor, color, and shape. It does not prevent spoilage of these foods.

Jams and Jellies with Reduced Sugar

Jellies and jams that contain modified pectin, gelatin, or gums may be made with noncaloric sweeteners. You can also make jams with less sugar than usual by using concentrated fruit pulp, which contains less liquid and less sugar. Two types of modified pectin are available for home use. One jells with one-third less sugar. The other is a low-methoxyl pectin that requires a source of calcium for jelling. To prevent spoilage, jars of these products may need to be processed longer in a boiling-water canner. Recipes and processing times provided with each modified pectin product must be followed carefully. The proportions of acids and fruits should not be altered, as spoilage may result. Acceptably jelled refrigerator fruit spreads may also be made with gelatin and sugar substitutes. Such products spoil at room temperature, must be refrigerated, and should be eaten within one month.

Frequent stirring and avoiding overcooking is key to producing successful jams and jellies.

Canning Meat

Fresh home-slaughtered red meats and poultry should be chilled and canned without delay. Do not can meat from sickly or diseased animals. Ice fish and seafoods after harvest; eviscerate immediately; and can them within two days. Here's an overview of what's involved in meat canning. (Check *USDA Guide 5: Preparing and Canning Poultry, Red Meats, and Seafoods* for specific recipes with processing times.)

Vegetable canning is well known; meat canning, less so. People raised on canned meat fondly recall its flavor and tenderness. Obtaining fresh meat from healthy animals is an important first step when canning meat.

Ground or Chopped Beef, Lamb, Pork, Sausage, Veal, or Venison

Choose fresh, chilled meat. With venison, add one part high-quality pork fat to three or four parts venison before grinding. Use freshly made sausage, seasoned with salt and cayenne pepper (sage may cause a bitter off-flavor). Shape chopped meat into patties or balls, or cut cased sausage into 3- to 4-inch links. Cook until lightly browned. Ground meat may be sautéed without shaping. Remove excess fat. Fill hot jars with pieces. Add boiling meat broth, tomato juice, or water, leaving a 1-inch headspace. Remove air bubbles, and adjust the headspace if needed. Add 1 teaspoon of salt per quart to the jars if desired. Wipe the rims of the jars with a dampened clean paper towel. Adjust the lids, and process the jars.

Strips, Cubes, or Chunks of Beef, Lamb, Pork, Veal, or Venison

Choose quality, chilled meat. Remove excess fat. Soak strong-flavored game meats for 1 hour in brine water containing 1 tablespoon of salt per quart; rinse. Remove large bones.

Hot Pack: Precook meat until rare by roasting, stewing, or browning in a small amount of fat. Add 1 teaspoon of salt per quart to the jar if desired. Fill hot jars with pieces, and add boiling broth, meat drippings, water, or tomato juice, leaving a 1-inch headspace. Remove air bubbles, and adjust the headspace if needed. Process the jars.

Raw Pack: Add 1 teaspoon of salt per quart to the jar if desired. Fill hot jars with raw meat pieces, leaving 1-inch headspace. Do not add liquid. Wipe rims of jars with a dampened clean paper towel. Adjust the lids, and process the jars.

Chicken or Rabbit

Choose freshly killed and dressed, healthy animals. Large chickens are more flavorful than fryers. Dressed chicken should be chilled for 6 to 12 hours before canning. Dressed rabbits should be soaked one hour in water containing 1 tablespoon of salt per quart, and then rinsed. Remove excess fat. Cut the chicken or rabbit into suitable sizes for canning. Can with or without bones.

Hot Pack: Boil, steam, or bake meat until about two-thirds done. Add 1 teaspoon of salt per quart to the jar if desired. Fill hot jars with pieces and hot broth, leaving a 1¼-inch headspace. Remove air bubbles, and adjust the headspace if needed.

Raw Pack: Add 1 teaspoon of salt per quart if desired. Fill hot jars loosely with raw meat pieces, leaving a 1¼-inch headspace. Do not add liquid. Wipe the rims of the jars with a dampened clean paper towel. Adjust the lids, and process the jars.

Meat Stock (Broth)

For beef stock, saw or crack fresh trimmed beef bones to extract extra flavor. For chicken and turkey stock, remove most of the meat from the carcass. To make stock, rinse the bones, and place them in a large stock pot or kettle; cover the bones with water; add the pot cover; and simmer three to four hours for beef, 45 minutes for poultry. When cool enough to handle, remove the bones; cool broth; and pick off the meat. Skim off fat; add meat trimmings removed from the bones to the broth; and reheat to boiling. Fill hot jars, leaving a 1-inch headspace. (See page 200 for more on headspace.) Wipe the rims of the jars with a dampened clean paper towel. Adjust the lids, and process the jars.

Having home-canned broth on hand gives you a head start on making soup.

Canning Fish

Use the following method for canning fish in pint jars, including blue fish, mackerel, salmon, steelhead trout, and other fatty fish except tuna. Always bleed and eviscerate fish immediately after catching, never more than 2 hours after they are caught. Keep cleaned fish on ice until you're ready to can it. Glasslike crystals of struvite, or magnesium ammonium phosphate, sometimes form in canned salmon. There is no way for the home canner to prevent these crystals from forming, but they usually dissolve when heated and are safe to eat.

If the fish is frozen, thaw it in the refrigerator before canning. Rinse the fish in cold water. You can add vinegar to the water (2 tablespoons per quart) to help remove slime. Remove the head, tail, fins, and scales; it is not necessary to remove the skin. You can leave the bones in most fish because the bones become very soft and are a good source of calcium. For halibut, remove the head, tail, fins, skin, and bones. Wash and remove all blood. Refrigerate all fish until you are ready to pack it in jars.

Fish being prepared for canning is often split lengthwise. Cut cleaned fish into 3½-inch lengths. If you have left the skin on the fish, pack the fish skin out for a nicer appearance or skin in for easier jar cleaning. Fill hot pint jars, leaving a 1-inch headspace. Add 1 teaspoon of salt per pint if desired. Do not add liquids. Carefully clean the jar rims with a clean, damp paper towel; wipe with a dry paper towel to remove any fish oil. Adjust the lids, and process the jars. Fish in half-pint or 12-ounce jars would be processed for the same amount of time as pint jars.

Making Sausage

Sausage making is not everyone's cup of tea, but should you raise and butcher livestock or go shares on a pig or cow, it's a great way to make thorough use of the carcass. Good sausage begins with good meat. Beef, veal, pork, lamb, mutton, and poultry are all suitable for use in sausage. If you slaughter your own animal, meat from the head, trimmings, and the thin cuts can be saved for sausage. Meat from the neck and back of poultry, and meat from the entire carcass of spent fowl are used. If you purchase meat, inexpensive cuts such as beef plates, chuck cuts, and pork jowls and shoulders can be used. Always use fresh, clean meat ingredients.

Venison and other game may be substituted for all or part of the lean meats in sausage recipes. Because wild game is slaughtered under less than desirable conditions, it is important to properly trim this type of meat. Be sure to remove any meat that is slimy, has an off-odor,

Sausage not only preserves meat, it puts to good use bits and pieces of the carcass that are difficult to deal with. Combined with vegetables and other fillers, marginal meat is made delicious.

or is dirty. Always keep meat cold. Avoid making the formula too lean, or you'll end up with sausage that is too dry and hard. Fresh pork sausage contains 30 to 45 percent fat. Smoked or roasted sausage contains 20 to 30 percent fat. Formulate the fat content just as you would the other ingredients in a sausage.

Curing Ingredients

Nitrate and nitrite are required to achieve the characteristic flavor, color, and stability of cured meat. During the curing process, nitrate and nitrite are converted to nitric oxide by microorganisms and combine with the meat pigment myoglobin to give the cured meat color. More importantly, nitrite provides protection against the growth of botulism-producing organisms, acts to retard rancidity, and stabilizes the flavor of the cured meat.

Exercise extreme caution when adding nitrate or nitrite to meat because too much of either of these ingredients can be toxic to humans. In using these materials, never use more than that called for in the recipe. A little is enough. Federal regulations permit a maximum addition of 2.75 ounces of sodium or potassium nitrate per 100 pounds of chopped meat, and 0.25 ounce of sodium or potassium nitrite per 100 pounds of chopped meat. Potassium nitrate

(saltpeter) was the salt historically used for curing. However, sodium nitrite alone, or in combination with nitrate, has largely replaced the straight nitrate cure used in the past.

Because these small quantities are difficult to weigh out on most available scales, be on the safe side and use a commercial premixed cure when nitrate or nitrite is called for in the recipe. The premixes have been diluted with salt so that the small quantities that must be added can more easily be weighed. This reduces the possibility of serious error in handling pure nitrate or nitrite. Many local grocery stores stock several brands of premix cure. Use this premix as the salt in the recipe, and it will supply the needed amount of nitrite simply and safely.

Remember, meats processed without nitrite are more susceptible to bacterial spoilage and flavor changes and probably should be frozen until used.

Seasonings

Salt is an essential ingredient in sausage. Salt is necessary for flavor, aids in preserving the sausage, and extracts the "soluble" meat protein at the surface of the meat particles. This protein is responsible for binding the sausage together when the sausage is heated and the protein coagulates. Most sausages contain 2- to 3-percent salt. Salt levels can be adjusted to your taste.

Seasonings and spices should be fresh. Most spices lose their natural flavor when held at room temperature for six months or more. For the best results, store seasonings at 55°F or below in airtight containers. Remember, the characteristic flavor of a sausage comes from the spices, herbs, and flavorings that are used, so buy the best you can find.

Commercial premixed seasonings are available for most sausages. For making small batches of sausage at home, premixed spices are excellent for providing fresh seasonings with good spice combinations.

Casings

Sausages may be formed into loaves and oven baked; however, most sausages are stuffed into casings. Natural casings are from sheep (¾-inch), hog (1¼ to 1¾ inches), and cattle (1¾ inches) intestines. These usually come in lengths of several feet packed in salt in 1-pound cups or in bulk by the yard. Although they add to the cost, they offer the advantage of being edible. One hank, or small container, of pork casings will stuff 40 to 50 pounds of sausage.

Edible synthetic casings made from collagen are also available in approximately the same sizes as the natural casings. Large synthetic casings that are used for slicing products, such as summer sausage or bologna, are not edible. These cellulose or fibrous casings have the advantage of being uniform in size (diameter) and generally free of defects. They are available in sizes from ¾ inch to 6 inches in diameter.

Be sure to select the proper size casing for the sausage being made. Small, edible natural casings from sheep or hogs are used for fresh sausage, while the larger beef casings are used for cooked and smoked sausages. Casings are readily available from online sausage-supply sites and from most butcher shops, meat-packing plants, and butcher-supply houses.

Other Sausage Additives

Many recipes call for holding the meat overnight to cure. This is required to allow the bacteria to convert the nitrite to nitric oxide. The addition of a reducing agent such as ascorbic acid (Vitamin C) speeds the curing reaction and eliminates the holding time. Another reducing agent, sodium erythrobate (isoascorbic acid) may also be used. Meat inspection regulations allow the use of ⅞ ounce per 100 pounds of meat.

Many sausages contain some additional ingredients called binders that may improve the flavor and help to retain the natural juiciness. Extenders like nonfat dry milk, cereal flours, and soy protein products bulk up the sausage ingredients. You may use these ingredients in most products, depending upon your taste.

Water is added to most sausage formulations to rehydrate the nonfat dry milk and to replace the expected moisture loss during smoking and cooking. Approximately 10 percent added water is used in most types of cooked sausage. A small amount of water (usually less than 3 percent) is added to fresh sausage to aid in stuffing, mixing, and processing. No water is added to sausages that are to be dried, such as summer sausage or pepperoni.

Small natural sausage casings are made from sheep intestines. Some synthetic casings are made of collagen and are edible; others are not edible but aid with slicing.

Temperatures

Meat products are extremely perishable and must be maintained under refrigeration (40°F or below). When you have finished processing a product, return it to the refrigerator. After the product has been formulated, smoke and cook the product to the required temperature, and then return the product to refrigeration. (See "Safe Storage," opposite.) Do not guess at the temperature of the product or rely on times stated in a recipe. Use a meat thermometer. Temperature-abused sausage can permit excessive microbial growth and result in product spoilage and food-borne illness.

Sanitation

In addition to maintaining sausage and all meat products at the proper temperature, it is also important to keep the work area sanitary. There is no substitute for keeping the tables, utensils, and ingredients clean and free from dirt and contamination to prevent problems later. Use plenty of hot water and soap before and after processing sausages. Always keep your hands clean. Be prepared to wash your hands and the utensils you will be using repeatedly while working, especially if you will be handling other items. These measures prevent spoilage and food-borne illness.

Sausage Making Steps

1. Weighing

Weighing or measuring the meat and spice ingredients is one of the most important steps in the preparation of a good sausage. Weigh the proper amounts of lean meat, fat, and each individual spice or added ingredient to be sure that the formulation is correct. There is nothing more disappointing than making sausage that is too hot or not properly seasoned. If it is not possible to weigh the ingredients, be sure to measure them carefully. Remember, weights are always more accurate than measures.

2. Mixing

Mixing the meat and other ingredients is a simple but important step. Before the spices and dry materials are added to the meat, cut the meat into 1- to 2-inch squares. Spread the meat in the bottom of a large pan. Sprinkle the spices and dry ingredients over the meat, and mix thoroughly. Add the water or wet ingredients last, and mix again. This mixing will ensure a uniform distribution of spices and develop the binding ability of the meat. If you will use nitrate or nitrite in the formula, dissolve it in a small amount of water before adding. This will ensure a uniform distribution throughout the sausage.

3. Grinding

Small manual or power grinders are available from most hardware or appliance stores. These are adequate for home sausage making. Larger models are available from restaurant and institutional suppliers.

Even a hand grinder with sharp, clean blades can produce ground meat and poultry (here) that is nicely textured.

Safe Storage

The length of time a sausage can be stored depends on the type of sausage. Fresh sausage is highly perishable and will only last 7 to 10 days. However, it may be frozen for 4 to 6 months if wrapped in moisture-vapor-proof wrap (freezer paper). Smoked sausages that have been cooked and contain salt and nitrite may last from 2 to 4 weeks under refrigeration. These types include smoked, Polish, cotto salami, and bologna. Summer sausages that have been fermented to produce the acid tangy flavor are more durable and may be stored for several weeks in the refrigerator.

The key to doing a good job grinding is to use sharp blades and plates that match. Clean out any gristle and bone fragments so that the plate and blade will fit together. Avoid smearing or crushing the meat through the plate. This will change the texture and color of the sausage, making it mushy. You may grind the sausage twice, especially if two meats, such as a fat meat and a lean meat, are being used. Grind each meat through a ⅜-inch coarse plate. Add the spices and other ingredients; mix them thoroughly; and then grind through the final ⅜- or ³⁄₁₆-inch plate. If two plates are not available, you can add the spices to the meat pieces and then grind it twice through the small plate. A single grind is usually not adequate.

4. Stuffing and Linking

Getting the sausage into the casings may present a problem if you do not have a stuffer. The simplest stuffer is a horn that fits the grinder; however, several other types of home stuffers may be used. Follow the manufacturer's instructions in assembling the stuffer. Slip the open end of the casing onto the horn. Put some meat through the stuffing horn until it is filled. Pull 1 or 2 inches of casing over the end of the horn, and knot it. Hold the mouth of the horn with your thumb and forefinger, and allow the casing to slide under your fingers as the casing fills. If you hold the casing too tightly, the pressure of the meat will tear the casing. Allow as few air pockets as possible to form in the casing. Prevent air pockets by packing the sausage tightly into the stuffer. Be sure the horn and casings are wet. This will allow the casings to feed freely off the horn. If the sausage is to be linked, you must stuff the casing loosely so that when twisted several turns to form the link, the casing will not burst. You can also make links by tying the casing with cotton string after stuffing.

Large casings (2 to 6 inches) are stuffed in the same manner as small casings. The casing should be pre-tied at one end and the open end fed completely over the horn. Grip the casing tightly with your thumb and forefinger on the mouth of the horn. Allow the casing to feed out as it fills, but avoid letting your fingers slip over the end of the horn. Leave about 2 inches of casing empty to tie off with a string. You can remove air pockets by pricking the casing with a pin. Tie the casing string tight, and leave enough string to hang the sausage in the smokehouse.

Drying Fruits and Vegetables

Food drying is one of the oldest methods of preserving food and is simple, safe, and easy to learn. Drying removes the moisture from the food, so bacteria, yeast, and mold cannot grow and spoil the food. Drying also slows down the action of enzymes (naturally occurring substances which cause foods to ripen) but does not inactivate them.

As drying removes moisture, the food shrinks and becomes lighter in weight. When the food is ready for use, water is added, and the food returns to its original shape. Foods can be dried in the sun, in an oven, or in a food dehydrator by using the right combination of warm temperature, low humidity, and air current. Warmth causes the moisture to evaporate. Low humidity allows moisture to move quickly from the food to the air. Air current speeds up drying by moving the surrounding moist air away from the food.

Sun Drying

The high sugar and acid contents of fruits make them safe to dry in the sun. Sun drying isn't recommended for vegetables and meats. Vegetables are low in sugar and acid, increasing the risk for food spoilage. Meats are high in protein, making them prone to microbial growth when heat and humidity cannot be controlled.

For sun drying, choose hot, dry, breezy days. Minimum temperature is 86°F—higher temperatures are better. It takes several days to dry foods outdoors and because the weather can be unpredictable, sun drying can be risky. Humidity below 60 percent, which is sometimes hard to find in humid summer weather, is best for sun drying. Unfortunately, these ideal conditions are often not available when fruit ripens.

Fruits dried in the sun are placed on trays made of screen or wooden dowels. Screens need to be safe for contact with food. The best screens are stainless steel, teflon-coated fiberglass, or plastic. Avoid screens made from "hardware cloth." This is galvanized metal cloth that is coated with cadmium or zinc. These materials can oxidize, leaving harmful residues on the food. Also avoid copper and aluminum screening. Copper destroys vitamin C and increases oxidation. Aluminum tends to discolor and corrode.

Most woods are fine for making trays. However, do not use green wood, pine, cedar, oak, or redwood, which can warp, stain the food, or cause off-flavors in the food.

Place trays on blocks to allow for better air movement around the food. Because the ground may be moist, it's best to place the racks or screens on a concrete driveway or patio. If possible, put a sheet of aluminum or sheet metal under the trays. The reflection of the sun on the metal increases the drying temperature. Cover the trays with cheesecloth to help protect the fruit from birds and insects. Fruits dried in the sun must be covered or brought under shelter at night. The cool night air condenses and could add moisture back into the food, thus slowing down the drying process. Various solar drying plans can be found online that focus sunlight and use convection to speed the drying time, reducing the risks of food spoilage or mold growth.

Sunlight and trays of food-safe mesh are all that is needed for drying these figs.

Pasteurization

Sun or solar-dried fruits and vine-dried beans need treatment to kill any insects and their eggs that might be on the food. Unless destroyed, the insects will eat the dried food. Here are two recommended pasteurization methods:

1. **Freezer Method:** Seal the food in freezer-type plastic bags. Place the bags in a freezer set at 0°F or below for at least 48 hours.
2. **Oven Method:** Place the food in a single layer on a tray or in a shallow pan. Place in an oven preheated to 160°F for 30 minutes. After either of these treatments the dried fruit is ready to be conditioned and stored.

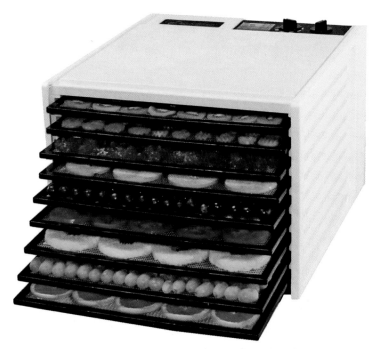

A factory-made dehydrator can be set at any temperature between 85°F and 140°F. This model comes with a timer that can be set to run the dryer for up 26 hours.

Drying Foods Indoors

Most foods can be dried indoors using modern dehydrators, convection ovens, or conventional ovens. A food dehydrator has an electric element for heat and a fan and vents for air circulation. Dehydrators are efficiently designed to dry foods quickly at 140°F. Twelve square feet of drying space dries about ½ bushel of produce.

By combining low heat, low humidity, and air flow, a standard kitchen oven can be used as a dehydrator. An oven is ideal for occasional drying of meat jerkies and fruit "leathers" (See page 217.) or for preserving excess produce like celery or mushrooms. Oven drying takes about two times longer than dehydrator drying because most ovens do not have built-in fans for the air movement. (However, convection ovens do have a fan.)

To use your oven, first check the dial to see whether it can register as low as 140°F. If your oven does not go this low, your food will cook and not dry. Use a thermometer to check the temperature at the "warm" setting. For air circulation, leave the oven door propped open 2 to 6 inches. Circulation can be improved by placing a fan outside the oven near the door. Because the door is left open, the temperature inside will vary. Adjust the door accordingly, and place an oven thermometer near the food.

Drying trays should be narrow enough to clear the sides of the oven and 3 to 4 inches shorter than the oven from front to back. Cake cooling racks placed on top of cookie sheets work well for some foods. The oven racks holding the trays should be 2 to 3 inches apart for air circulation.

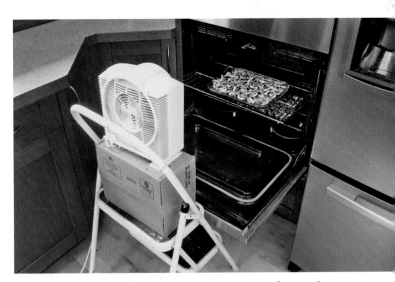

Before attempting to dry produce in your oven, make sure it can consistently register as low a temperature as 140°F. This is the typical setup for oven drying. To get started, close the door, keeping it open 2 to 6 in.

213

Drying Fruits

Dried fruits are unique, tasty, and nutritious. Begin by washing the fruit and coring it if needed. For drying, fruits can be cut in half or sliced. Some can be left whole.

Thin, uniform, peeled slices dry the fastest. You can leave the peel on the fruit, but unpeeled fruit takes longer to dry. You can core and slice apples in rings, wedges, or chips. Slice bananas in coins, chips, or sticks. Fruits dried whole take the longest to dry. Before drying, you need to "check" or crack the skins to speed drying. To "check" the fruit, place it in boiling water and then in cold water.

Because fruit contains sugar and is sticky, spray the drying trays with nonstick cooking spray before placing the fruit on the trays. After the fruit dries for one to two hours, lift each piece gently with a spatula and turn.

Uniformly thin wedges of apple dry well. Removing the skin from fruit speeds up the drying process even more.

Pretreating Fruit

Pretreatments prevent fruits from darkening. Many light-colored fruits, such as apples, darken rapidly when cut and exposed to air. If not pretreated, these fruits will continue to darken after they have dried.

For long-term storage of dried fruit, sulfuring or using a sulfite dip are the best pretreatments. However, sulfites found in the food after either of these treatments has been found to cause asthmatic reactions in a small portion of the asthmatic population. Thus, some people may want to use the alternative shorter-term pretreatments. If home-dried foods are eaten within a short time, there may be little difference in the long- and short-term pretreatments.

Sulfuring: This old method of pretreating fruits involves lighting sublimed sulfur and letting it burn in an enclosed box with the fruit. The sulfur fumes penetrate the fruit and act as a pretreatment by retarding spoilage and darkening of the fruit. Fruits must be sulfured out-of-doors where there is adequate air circulation. (For more information contact your County Extension Office.)

Sulfite Dip: Sulfite can achieve the same long-term antidarkening effect as sulfuring but more quickly and easily. Use sodium bisulfite, sodium sulfite, or sodium metabisulfite that is USP (food grade) or Reagant grade (pure). To purchase, check online or local sources where wine-making supplies are sold.

For sulfite dip, dissolve ¾ to 1½ teaspoons of sodium bisulfite per quart of water. (If using sodium sulfite, use 1½ to 3 teaspoons. If using sodium metabisulfite, use 1 to 2 tablespoons.) Place the prepared fruit in the mixture, and soak 5 minutes for slices, 15 minutes for halves. Remove fruit; rinse lightly under cold water; and place on drying trays. You can dry sulfited foods indoors or outdoors. (This solution can be used only once. Make a new one for the next batch.)

Ascorbic Acid: Ascorbic acid (vitamin C) mixed with water is a safe way to prevent fruit browning. However, its protection does not last as long as sulfuring or sulfiting. Ascorbic acid is available in powdered and tablet form, from drugstores and grocery stores. One teaspoon of powdered ascorbic acid is equal to 3,000 mg of ascorbic acid in tablet form. (If you buy 500 mg tablets, this equals six tablets.) Mix 1 teaspoon of powdered ascorbic acid (or 3,000 mg of ascorbic acid tablets, crushed) in 2 cups of water. Place the fruit in the solution for 3 to 5 minutes. Remove the fruit; drain it well; and place it on dryer trays. After you use this solution twice, add more acid.

Drying the Prepared Fruit

Whichever drying method you choose—sun drying, oven drying, or dehydrator drying—place the fruit in a single layer on the drying trays. The pieces should not touch or overlap. Follow the directions for the drying method you choose, and dry until the food tests dry. Food dries much faster at the end of the drying period, so watch it closely.

Because dried fruits are generally eaten without being rehydrated, you should not dehydrate them to the point of brittleness. Most fruits should have about 20 percent moisture content when dried.

To test for dryness, cut several cooled pieces in half. There should be no visible moisture, and you should not be able to squeeze any moisture from the fruit. Some fruits may remain pliable but shouldn't be sticky or tacky. When folded in half, a piece shouldn't stick to itself. Berries should be dried until they rattle when shaken.

After drying, cool fruit 30 to 60 minutes before packaging. Packaging food warm can lead to sweating and moisture buildup. However, excessive delays in packaging could allow moisture to re-enter food. Remember, if you have dried fruit in the sun, you must pasteurize it before you package it.

Ascorbic Acid Mixtures: Ascorbic acid mixtures are a mixture of ascorbic acid and sugar sold for use on fresh fruits and in canning or freezing. It is more expensive and not as effective as using pure ascorbic acid. Mix 1½ tablespoons of ascorbic acid mixture with 1 quart of water. Place the fruit in the mixture and soak 3 to 5 minutes. Drain the fruit well, and place it on dryer trays. After you use this solution twice, add more ascorbic acid mixture.

Fruit Juice Dip: A fruit juice that is high in vitamin C can also be used as a pretreatment, though it is not as effective as pure ascorbic acid. Juices high in vitamin C include orange, lemon, pineapple, grape, and cranberry. Each juice adds its own color and flavor to the fruit. To make a fruit juice dip, place enough juice to cover fruit in a bowl. Add cut fruit. Soak 3 to 5 minutes; remove the fruit; drain it well; and place it on dryer trays. You may use this solution twice before replacing it. (You can consume the used juice.)

Honey Dip: Many store-bought dried fruits have been dipped in a honey solution. You can make a similar dip at home. Honey-dipped fruit is much higher in calories. Mix ½ cup of sugar with 1½ cups of boiling water. Cool to lukewarm, and add ½ cup of honey.

Place the fruit in the dip, and soak it for 3 to 5 minutes. Remove it; drain it well; and place on dryer trays.

Syrup Blanching: Blanching fruit in syrup helps it retain color fairly well during drying and storage. The resulting product is similar to candied fruit. Fruits that can be syrup blanched include apples, apricots, figs, nectarines, peaches, pears, and plums. Combine 1 cup of sugar, 1 cup of light corn syrup, and 2 cups of water in a saucepot. Bring it to a boil. Add 1 pound of prepared fruit, and simmer it for 10 minutes. Remove it from the heat, and let the fruit stand in hot syrup for 30 minutes. Lift the fruit out of the syrup; rinse it lightly in cold water; drain it on paper towels; and place it on dryer trays.

Steam Blanching: Steam blanching also helps retain color and slow oxidation. However, the flavor and texture of the fruit is changed. Place several inches of water in a large saucepot with a tight-fitting lid. Heat to boiling. Place the fruit no more than 2 inches deep in a steamer pan or wire basket over boiling water. Cover it tightly with a lid, and begin timing immediately. Consult your recipe for the blanching time. Check for even blanching halfway through the blanching time. Some fruit may need to be stirred. When done, remove excess moisture using paper towels, and place the fruit on dryer trays.

Fruit-by-Fruit Drying Steps

Apples: Select mature, firm apples. Wash it well. Pare, if desired, and core. Some peeled fruits dry faster than unpeeled fruits. Cut in rings or slices ⅛ to ¼ inch thick, or cut in quarters or eighths. Dip in ascorbic acid or other antidarkening/antimicrobial solution for 10 minutes. Remove from solution, and drain well. Arrange in a single layer on trays, pit side up. Dry until soft, pliable, and leathery, with no moist area in the center when cut.

Apricots: Select firm, fully ripe fruit. Wash it well. Cut in half, and remove the pit. Do not peel. Dip in ascorbic acid or other antidarkening/antimicrobial solution for 10 minutes. Remove from solution and drain well. Arrange in a single layer on trays, pit side up, with the cavity popped up to expose more flesh to the air. Dry until soft, pliable, and leathery, with no moist area in the center when cut.

Bananas: Select firm, ripe fruit. Peel and cut it in ⅛-inch slices. Dip in ascorbic acid or other solution for 10 minutes. Remove, and drain well. Arrange in a single layer on trays. Dry until tough and leathery.

Berries: Select firm, ripe fruit. Wash it well. Leave whole or cut in half. Dip in boiling water 30 seconds to crack skins, or dip in ascorbic acid or other antimicrobial solution for 10 minutes. Remove, and drain well. Arrange on drying trays no more than two berries deep. Dry until hard and berries rattle when you shake the trays.

Cherries: Select fully ripe fruit. Wash it well. Remove stems and pits. Dip whole cherries in boiling water for 30 seconds; then plunge cherries into ice water to crack skins. May also dip in ascorbic acid or other antimicrobial solution for 10 minutes. Remove, and drain well. Arrange in a single layer on trays. Dry until tough, leathery, and slightly sticky.

Citrus Peel: Select thick-skinned oranges with no signs of mold or decay and no color added to skin. Scrub oranges well with a brush under cool, running water. Thinly peel outer ¹⁄₁₆ to ⅛ inch of the peel; avoid the white bitter part. Dip in ascorbic acid or other antidarkening/antimicrobial solution for 10 minutes. Remove from solution, and drain well. Arrange in single layers on trays. Dry at 130°F for 1 to 2 hours; then at 120°F until crisp.

Figs: Select fully ripe fruit. Wash or clean it well with damp towel. Peel dark-skin varieties if desired. Leave whole if small or partly dried on tree; cut large figs in half or in slices. If drying whole figs, crack skins by dipping in boiling water for 30 seconds, then plunging them into ice water. For cut figs, dip in ascorbic acid or other antimicrobial solution for 10 minutes. Remove, and drain well. Arrange in single layers on trays. Dry until leathery and pliable.

Grapes and Black Currants: Select seedless varieties. Wash fruit, and remove stems. Cut in half, or leave whole. If drying whole, crack skins by dipping in boiling water for 30 seconds, then plunging them into ice water. If halved, dip in ascorbic acid or other antimicrobial solution for 10 minutes. Remove, and drain well. Dry until pliable and leathery, with no moist area in the center.

Melons: Select mature, firm fruits that are heavy for their size; cantaloupe dries better than watermelon. Scrub the outer surface well with a brush under cool, running water. Remove the outer skin, any fibrous tissue, and seeds. Cut into ¼- to ½-inch-thick slices. Dip in ascorbic acid or other antimicrobial solution for 10 minutes. Remove, and drain well. Arrange in a single layer on trays. Dry until leathery and pliable, with no pockets of moisture.

Nectarines and Peaches: Select ripe, firm fruit. Wash and peel it. Cut in half, and remove the pit. Cut in quarters or slices if desired. Dip in ascorbic acid or other antidarkening/antimicrobial solution for 10 minutes. Remove, and drain well. Arrange in a single layer on trays, pit side up. Turn halves over when visible juice disappears. Dry until leathery and somewhat pliable.

Pears: Select ripe, firm fruit. Bartlett variety is recommended. Wash the fruit well. Remove the skin of the fruit if desired. Cut in half lengthwise, and core. Cut in quarters or eighths, or slice ⅛- to ¼-inch thick. Dip in ascorbic acid or other antidarkening/antimicrobial solution for 10 minutes. Remove the fruit and drain. Arrange in a single layer on trays, pit side up. Dry until springy and suedelike, with no pockets of moisture.

Plums: Wash the fruit well. Leave whole if small; cut large fruit into halves (pit removed) or slices. If left whole, crack skins in boiling water 1 to 2 minutes. If cut in half, dip in ascorbic acid or other antimicrobial solution for 10 minutes. Remove the fruit, and drain. Arrange in a single layer on trays, pit side up, cavity popped out. Dry until pliable and leathery; the pit should not slip when squeezed if the prune is not cut.

Conditioning Fruits

When dried fruit is taken from the dehydrator or oven, the remaining moisture may not be distributed equally among the pieces because of their size or their location in the dehydrator. Conditioning is a process used to equalize the moisture and reduce the risk of mold growth.

To condition the fruit, take the dried fruit that has cooled and pack it loosely in plastic or glass jars. Seal the containers, and let them stand for 7 to 10 days. The excess moisture in some pieces will be absorbed by the drier pieces. Shake the jars daily to separate the pieces and check for moisture condensation. If condensation develops in the jar, return the fruit to the dehydrator for more drying. After conditioning, package and store the fruit.

Making Fruit "Leather"

Fruit "leather" is a tasty, chewy, dried fruit product. Fruit leather is made by pouring pureed fruit onto a flat surface for drying. When the fruit is dried, you pull it from the surface and roll it. It gets the name *leather* from the fact that when pureed fruit is dried, it is shiny and has the texture of leather. Here's how to make leather from fresh fruit:

1. Select ripe or slightly overripe fruit.
2. Wash fresh fruit or berries in cool water. Remove peels, seeds, and stems.
3. Cut the fruit into chunks. Use 2 cups of fruit for each 13- x 15-inch fruit leather. Puree fruit until smooth.
4. Add 2 teaspoons of lemon juice or ⅛ teaspoon of ascorbic acid (375 mg) for each 2 cups of light-colored fruit to prevent darkening. Optional: to sweeten, add corn syrup, honey, or sugar. Corn syrup or honey is best for longer storage because it prevents crystals. Sugar is fine for immediate use or short storage. Use ¼ to ½ cup of sugar, corn syrup, or honey for each 2 cups of fruit. Saccharin-based sweeteners can also be used to reduce tartness without adding calories. Aspartame sweeteners may lose sweetness during drying.

For a tasty way to preserve fruit, make fruit "leather." Puree the fruit; then pour it on a tray for drying. The resulting leather is a flavorful, chewy snack.

You can also use canned or frozen fruits, including fruit you canned yourself. To do so, drain the fruit and save the liquid. Use the same measurements and additives as you would for fresh fruit. If the puree is too thick, use the reserved liquid to thin it. Applesauce can be dried alone or added to any fruit puree as an extender. It decreases tartness and makes the leather smoother and more pliable.

For drying leather in the oven or sun, line cookie sheets with plastic wrap. In a dehydrator, use plastic wrap or the specially designed plastic sheets that come with the dehydrator. Pour the puree onto the lined cookie sheets or tray. Spread it evenly to a thickness of ⅛ inch.

Dry the fruit leather at 140°F until no indention is left when you touch the center with your finger. This could take about 6 to 8 hours in the dehydrator, up to 18 hours in the oven, and 1 to 2 days in the sun. While it's still warm, peel it from the plastic wrap. Cool, and rewrap the leather in plastic to store.

Drying Vegetables

Because vegetables contain less acid than fruits, they must be dried until they are brittle. To prepare vegetables for drying, wash in cool water to remove soil and chemical residues. Prepare vegetables according to the directions in the table opposite. Keep pieces uniform in size so that they'll dry at the same rate. Prepare only as many as can be dried at one time.

Pretreating Vegetables

Blanching is a necessary step in preparing vegetables for drying. Blanching stops the enzyme action that could cause loss of color and flavor during drying and storage. It also shortens the drying and rehydration time by relaxing the tissue walls.

Water Blanching: Fill a large pot two-thirds full of water; cover it; and bring it to a rolling boil. Place the vegetables in a wire basket or a colander, and submerge them in the water. Cover and blanch according to directions. (See opposite.) Begin timing when the water returns to boiling. If it takes longer than 1 minute for the water to come back to boiling, you added too many vegetables. Reduce the amount in the next batch.

Steam Blanching: Use a deep pot with a tight-fitting lid and a wire basket, colander, or sieve placed so that the steam will circulate freely around the vegetables. Add water to the pot, and bring it to a rolling boil. Place the vegetables loosely in the basket no more than 2 inches deep. Make sure the water does not come in contact with the vegetables. Cover the pot, and steam according to the directions. (See opposite.)

Cooling and Drying the Prepared Vegetables

After blanching, dip the vegetables briefly in cold water. When they feel only slightly hot to the touch, drain the vegetables by pouring them directly onto the drying tray held over the sink. Arrange the vegetables in a single layer. Then place the tray immediately in the dehydrator or oven. Watch the vegetables closely. They dry much more quickly at the end of the process and could scorch.

Vegetables should be dried until they are brittle or crisp. Some vegetables would actually shatter if hit with a hammer. At this stage, they should contain about 10-percent moisture. Because they are so dry, they do not need conditioning like fruits.

Drying Beans

Dried beans have been a farm staple for millennia because they store well and when combined with corn, rice, or other grains, form a complete protein. Popular dry beans include kidney, navy, pinto, black, and fava, all of which can be kept for a year or more if handled correctly.

When bean pods turn brown and dry but before they split open, the beans are ready for harvest. Sometimes the beans will rattle inside the pod. Uproot the plants, and hang them upside down for two to three days to complete the drying. Test the dryness of the beans by biting one. You should barely be able to dent the flesh. If they seem soft, dry them some more.

Thresh by squeezing the pod and breaking out the beans. Shake them on a coarse screen to remove any dirt and debris. Next, freeze them for at least 4 hours to kill any insect larvae. Store them in airtight containers in a cool, dry area away from sunlight. Beans are sometimes soaked before cooking, though many bean aficionados think this step steals from their flavor.

Steps for Drying Vegetables

*Blanching times are for 3,000 to 5,000 feet above sea level. Times will be slightly shorter for lower altitudes and slightly longer for higher altitudes or for large quantities of vegetables.
**WARNING: the toxins of poisonous varieties of mushrooms are not destroyed by drying or by cooking. Only an expert can differentiate between poisonous and edible varieties.

VEGETABLE	PREPARATION	BLANCHING TIME* (MINS.)	DRYING TIME (HRS.)	DRYNESS TEST
Asparagus	Wash thoroughly. Halve large tips.	4–5	6–10	Leathery to brittle
Beans, green	Wash. Cut in pieces or strips.	4	8–14	Very dry, brittle
Beets	Cook as usual. Cool; peel. Cut into shoestring strips ⅛ in. thick.	None	10–12	Brittle, dark red
Broccoli	Wash. Trim; cut as for serving. Quarter stalks lengthwise.	4	12–15	Crisp, brittle
Brussels sprouts	Wash. Cut in half lengthwise through stem.	5–6	12–18	Tough to brittle
Cabbage	Wash. Remove outer leaves; quarter; and core. Cut into strips ⅛ in. thick.	4	10–12	Crisp, brittle
Carrots, parsnips	Use only crisp, tender vegetables. Wash. Cut off roots and tops; peel. Cut in slices or strips ⅛ in. thick.	4	6–10	Tough to brittle
Cauliflower	Wash. Trim; cut into small pieces.	4–5	12–15	Tough to brittle
Celery	Trim stalks. Wash stalks and leaves thoroughly. Slice stalks.	4	10–16	Very brittle
Chili peppers, green	Wash. To loosen skins, cut slit in skin, then rotate over flame 6–8 minutes or scald in boiling water. Peel and split pods. Remove seeds and stem. (Wear gloves if necessary.)	None	12–24	Crisp, brittle, medium green
Chili peppers, red	Wash thoroughly. Slice, or leave whole if small.	4	12–24	Shrunken, dark-red pods, flexible
Corn, cut	Husk; trim. Wash well. Blanch until milk in corn is set. Cut kernels from the cob.	4–6	6–10	Crisp, brittle
Eggplant	Wash; trim; cut into ¼-in. slices.	4	12–14	Leathery to brittle
Horseradish	Wash; remove small rootlets and stubs. Peel or scrape roots. Grate.	None	6–10	Brittle, powdery
Mushrooms**	Scrub. Discard tough, woody stalks. Slice tender stalks ¼ in. thick. Peel large mushrooms; slice. Leave small mushrooms whole.	None	8–12	Dry and leathery
Okra	Wash thoroughly. Cut into ½-in. pieces, or split lengthwise.	4	8–10	Tough, brittle
Onions	Wash; remove outer paper skin. Remove tops and root ends; slice ⅛ to ¼ in. thick.	4	6–10	Very brittle
Parsley; other herbs	Wash thoroughly. Separate clusters. Discard long or tough stems.	4	4–6	Flaky
Peas	Shell, and wash.	4	8–10	Hard, wrinkled, green
Peppers; pimentos	Wash; stem. Remove core and seeds. Cut into ¼- to ½-in. strips or rings.	4	8–12	Tough to brittle
Potatoes	Wash; peel. Cut into ¼-in. shoestring strips or ⅛-in.-thick slices.	7	6–10	Brittle
Spinach; greens like kale, chard, mustard	Trim, and wash very thoroughly. Shake or pat dry to remove excess moisture.	4	6–10	Crisp
Squash, summer or banana	Wash; trim; cut into ¼-in. slices.	4	10–16	Leathery to brittle
Squash, winter	Wash rind. Cut into pieces. Remove seeds and cavity pulp. Cut into 1-in.-wide strips. Peel rind. Cut strips crosswise into pieces about ⅛ in. thick.	4	10–16	Tough to brittle
Tomatoes	Steam or dip in boiling water to loosen skins. Chill in cold water. Peel. Slice ½ in. thick, or cut in ¾-in. sections. Dip in solution of 1 tsp. citric acid/quart water for 10 minutes.	None	6–24	Crisp

7 Harvest Home

Packaging and Storing Dried Foods

After foods are dried, cool them completely. Then package them in clean moisture-vapor-resistant containers. Glass jars, metal cans, or freezer containers are good storage containers if they have tight-fitting lids. Plastic freezer bags are acceptable, but they are not insect and rodent proof. Fruit that has been sulfured or sulfited should not touch metal. Place the fruit in a plastic bag before storing it in a metal can.

Dried food should be stored in a cool, dry, dark place. Most dried fruits can be stored for one year at 60°F, six months at 80°F. Dried vegetables have about one-half the shelf life of fruits. To store any dried product longer, place it in the freezer. You can eat dried fruits as is or reconstitute them. You must reconstitute dried vegetables. Eat fruit leathers and meat jerky as is.

To reconstitute dried fruits or vegetables, add water to the fruit or vegetable, and soak until the desired volume is restored. (See the table on rehydrating dried food, below.) Do not soak the food for too long.

For soups and stews, add the dehydrated vegetables, without rehydrating them. They will rehydrate as the soup or stew cooks. Also, leafy vegetables and tomatoes do not need soaking. Add enough water to cover them, and simmer until tender. **CAUTION:** if soaking takes more than 2 hours, refrigerate the product for the remainder of the time.

Rehydrating Dried Foods

PRODUCT	WATER TO ADD TO 1 CUP DRIED FOOD (CUPS)	MINIMUM SOAKING TIME (HOURS)
FRUITS	(Water is at room temperature)	
Apples	1½	½
Pears	1¾	1¼
Peaches	2	1¼
VEGETABLES	(Boiling water)	
Asparagus	2¼	1½
Beans, lima	2½	1½
Beans, green snap	2½	1
Beets	2¾	1½
Carrots	2¼	1
Cabbage	3	1
Corn	2¼	½
Okra	3	½
Onions	2	¾
Peas	2½	½
Pumpkin	3	1
Squash	1¾	1
Spinach	1	½
Sweet Potatoes	1½	½
Greens	1	¾

Freezing Your Harvest

Freezing the produce and meat you raise is one of the simplest and safest preservation methods. A small chest-type freezer can cost less than $200, and if Energy Star rated, it will save you 15 to 30 percent in energy costs over older freezers. To protect your harvest from the dry climate of the freezer, use the right packing materials. In general, packaging materials must be

- **Moisture-vapor resistant.**
- **Durable and leakproof.**
- **Able to withstand low temperatures so that they don't become brittle and crack.**
- **Resistant to oil, grease, and water.**
- **Strong enough to protect foods from absorption of off-flavors or -odors.**
- **Easy to seal and easy to mark.**

Small is Better. Do not freeze produce in containers larger than ½ gallon. Foods in larger containers freeze too slowly to result in a satisfactory product.

Rigid Containers

Rigid containers made of plastic or glass are suitable for all packs and are especially good for liquid packs. Straight sides on rigid containers make the frozen food much easier to get out. Rigid containers are often reusable and make the stacking of foods in the freezer easier. Cardboard cartons for cottage cheese, ice cream, and milk are not sufficiently moisture-vapor resistant to be suitable for long-term freezer storage.

If you're using glass jars, choose wide-mouth, dual-purpose jars made for freezing and canning; regular glass jars break easily at freezer temperatures. The wide mouth allows easy removal of partially thawed foods. If standard canning jars (those with narrow mouths) are used for freezing, leave extra headspace to allow for expansion of foods during freezing. Expansion of the liquid could cause the jars to break at the neck. Some foods will need to be thawed completely before removal from the jar.

Covers for rigid containers should fit tightly. If they do not, reinforce the seal with freezer tape. Freezer tape sticks at freezing temperatures.

Bags or Wrappings

Flexible freezer bags and moisture-vapor-resistant wrapping materials such as plastic freezer wrap, freezer paper, and heavy-duty aluminum foil are suitable for dry packed products with little or no liquid. Bags and wraps work well for foods with irregular shapes. Bags can also be used for liquid packs. (A liquid pack is fruit packed in juice, sugar, syrup, or water, or crushed or pureed fruit. A dry pack is fruit or vegetables packed without added sugar or liquid.)

Plastic freezer bags are available in a variety of sizes. There are two types of closures. One type is twisted at the top, folded over, and wrapped with twist ties included in the package. The other is zipped or pressed to seal a plastic channel. Press to remove as much air as possible before closing.

To completely remove air and greatly reduce the chance of freezer burn, consider a vacuum bagger. This handy machine draws the air out of the bagging material, then seals it, shrink-wrapping the food. One advantage is ease of defrosting: the sealed bag can be set in water without fear of contamination.

Costing about $70, a vacuum bagger extends the freezer life of frozen foods and eases defrosting.

Best Storage Time

Freezing cannot improve the flavor or texture of any food, but when properly done it can preserve most of its fresh quality. The storage times listed below are approximate months of storage for some food products, assuming the food has been prepared and packaged correctly and stored in the freezer at or below 0°F. For best quality, use the shorter storage times. If stored longer than these recommended times, the food should still be safe, just lower in quality.

Flavor Frozen Out

Most items retain their original flavors after being frozen, but a few are altered. When using seasonings and spices, season lightly before freezing, and add additional seasonings when reheating or serving. Here are some changes to watch for:

- **Pepper, cloves, garlic, green pepper, imitation vanilla, and some herbs tend to develop a strong flavor and become bitter.**
- **Onions tend to break down and lose some of their texture during freezing.**
- **Celery seasonings become stronger.**
- **Curry develops a musty off-flavor.**
- **Salt loses flavor and has the tendency to increase rancidity of any item containing fat.**

Aside from the type of food, these factors influence the shelf life of frozen items:

Was the food properly blanched?

- Blanching stops enzyme actions that cause loss of flavor, color, texture, and nutrients even in frozen storage. (See "Pretreating Vegetables," page 218.)

Was the food packaged in appropriate materials?

- Allow enough headspace so that food can expand without breaking package seals.

Freezing Headspace

Here's the headspace to allow between packed food and the container when freezing. See page 200 for more on headspace.

TYPE OF PACK	Container with wide top opening		Container with narrow top opening	
	PINT	QUART	PINT	QUART
Liquid Pack*	½ in.	1 in.	¾ in.	1½ in.
Dry Pack**	½ in.	½ in.	½ in.	½ in.
Juices	½ in.	1 in.	1½ in.	1½ in.

*Fruit packed in juice, sugar, syrup, or water; crushed or pureed fruit.

**Fruit or vegetable packed without added sugar or liquid.

Storing Frozen Food

FOOD	APPROXIMATE MONTHS OF STORAGE AT 0°F
Fruits and Vegetables	8–12
Poultry	6–9
Fish	3–6
Ground Meat	3–4
Cured or Processed Meat	1–2

- Label each package; include the name of the product, any added ingredients, and the packaging date.
- Use freezer tape, or pens and labels made especially for freezer use.

Was the food stored at an acceptable temperature?

- Freeze and store food at 0°F or lower.
- Freeze foods as soon as you pack and seal them.
- Do not overload the freezer with unfrozen food.
- Leave space among new, warm packages so that the cold air can circulate freely around them. When the food is frozen, stack and store the packages close together.

A good policy to follow is "first in, first out." Rotate foods so that you use the older items first and enjoy your food at its best quality.

Freezing Pointers

- Freeze foods at 0°F or lower. To facilitate more rapid freezing, set the temperature control of the freezer at -10°F or lower about 24 hours in advance of loading fresh food.
- Freeze foods as soon you pack and seal them.
- Do not overload your freezer with unfrozen food. Add only the amount that will freeze within 24 hours, which is usually 2 to 3 pounds of food per cubic foot of storage space. Overloading slows down the freezing rate, and foods that freeze too slowly may lose quality.
- Place packages in contact with refrigerated surfaces in the coldest part of the freezer. Leave a little space between packages so that air can circulate freely. When the food is frozen, store the packages close together.

Packaging and Labeling Foods

- Cool all foods and syrup before packaging. This speeds up freezing and helps retain the natural color, flavor, and texture of food.
- Pack foods in quantities that will be used for a single meal.
- Follow directions for each individual food to determine which can be packed dry and which need added liquid. Some loose foods, such as blueberries, may be "tray packed," placed in a single layer on a tray and frozen. Once frozen, the food can be placed in a standard storage container where each piece will remain separate.
- Pack foods tightly, leaving as little air as possible in the package.
- Most foods require headspace between the packed food and closure to allow for expansion of the food as it freezes. Foods that are exceptions and do not need headspace include loose-packed vegetables such as asparagus and broccoli, bony pieces of meat, tray-packed foods, and breads.
- Seal rigid containers carefully. Use a tight lid, and keep the sealing edge free from moisture or food to ensure a good closure. Secure loose-fitting covers using freezer tape.
- Label each package, including the name of the product, any added ingredients, packaging date, the number of servings, and the form of the food, such as whole, sliced, etc. Use freezer tape, marking pens or crayons, or gummed labels that are made especially for freezer use.

Safe Thawing

Food must be kept at a safe temperature during defrosting. Foods are safe indefinitely while frozen; however, as soon as food begins to defrost and become warmer than 40°F, any bacteria that may have been present before freezing can begin to multiply. Never thaw food at room temperature or in warm water. Even though the center of a package may still be frozen as it thaws on the counter or in the warm water, the outer layer of the food is in the "Danger Zone," between 40 and 140°F. These are temperatures where bacteria multiply rapidly.

Thaw food in the refrigerator at 40°F or less, in cold, running water less than 70°F, or in the microwave if you'll be cooking or serving it immediately.

Thawing in cold water requires less time but more attention than thawing in the refrigerator. This should only be used if the water is kept cold (less than 70°F) and the food will thaw in under 2 hours. The food must be in a leak-proof package or plastic bag. As an alternative to constantly running water, the bag of food could be submerged in cold tap water, changing the water every 30 minutes as the food continues to thaw.

Thawing in the microwave oven produces some uneven heating patterns. Some parts of a food may actually start to cook before other sections completely thaw. Holding partially cooked food is not recommended because any bacteria present wouldn't have been destroyed and, indeed, may have reached optimal temperatures for bacteria to grow. Use the microwave when the food will be cooked immediately after thawing.

Using Frozen Fruit

When serving frozen fruits for dessert, serve them while there are still a few ice crystals in the fruit. This helps compensate for the mushy texture frozen fruits have when thawed.

Frozen fruit in the package can be thawed in the refrigerator, under running water, or in a microwave oven if thawed immediately before use. Turn the package several times for more even thawing. Allow 6 to 8 hours in the refrigerator for thawing a 1-pound package of fruit packed in syrup. Allow ½ to 1 hour for thawing in running cool water.

Fruit packed with dry sugar thaws slightly faster than that packed in syrup. Both sugar and syrup packs thaw faster than unsweetened packs.

Thaw only as much as you need at one time. If you have leftover thawed fruit, it will keep better if you cook it. To cook, first thaw fruits until pieces can be loosened; then cook as you would cook fresh fruit. If there is not enough juice to prevent scorching, add water as needed.

When using frozen fruits in cooking, allow for any sugar added at the time of freezing. Frozen fruits often have more juice than called for in recipes for baked products using fresh fruits. In that case, use only part of the juice or add more thickening for the extra juice.

Using Frozen Vegetables

Most frozen vegetables should be cooked without thawing first. Partially thaw corn on the cob before cooking in order for the cob to be heated through by the time the corn is cooked. Letting the corn sit after thawing or cooking causes sogginess. Leafy greens, such as turnip greens and spinach, cook more evenly if you partially thaw them before cooking.

To cook, bring water to a boil in a covered saucepan. The amount of water needed depends on the vegetable and the size of the package. It is important to use as little water as possible because some nutrients dissolve into the water. For most vegetables, ½ cup of water is enough for a pint package. Any frost in the package adds moisture.

Place the frozen vegetables in boiling water; cover the pan; and bring the water quickly back to a boil. To ensure uniform cooking, it may be necessary to separate pieces care-

fully using a fork. When the water is boiling throughout the pan, reduce the heat and cook until done. Be sure the pan is covered to keep in the steam, which aids in cooking. Cook gently until vegetables are just tender. Add seasonings as desired, and serve immediately or use in casseroles.

Uses for Frozen Fruit

- Frozen fruits can be used the same as fresh fruits in preparing pies, upside-down cakes, sherbets, ices, and salads. Some fruits, especially boysenberries, make better jellies when frozen than when fresh because freezing and thawing causes the juices to be released from the cells and the natural fruit color dissolves in the juice.
- Serve crushed fruit the same as raw fruit after it is partially or completely thawed; use it after thawing as a topping for ice cream or cake or a filling for sweet rolls or for jam.
- Use thawed purees in puddings, ice cream, sherbets, jams, pies, ripple cakes, fruit-filled coffee cakes, and rolls.
- Use frozen fruit juice as a beverage after it is thawed but while it is still cold. Some juices, such as sour cherry, plum, grape, and berry, can be diluted 3:1 to 2:1 with water or a bland juice.

Eggs, Butter, Milk, and Cheese

If you are keeping chickens or milking goats, at times you may find yourself with too much produce to handle. Rest easy: eggs, butter, milk, cream, and cheese can be frozen. To freeze eggs, crack them into a bowl and stir to combine the albumen and yolk. Try not to make bubbles as you pour the eggs into the freezing container. Thaw eggs, butter, milk, and cheese in the refrigerator. Before using the thawed cream, mix or blend it slightly.

Meat, Fish, and Poultry

Meat, fish, and poultry can be cooked from the frozen or thawed stage. Frozen meats, fish, and poultry are best when thawed in the refrigerator in their original wrappings. For faster thawing, place the meat or fish in waterproof wrapping in cold, slowly running water. If you can't keep water running slowly over the package, place it in a large container of cold water. Change the water at least every 30 minutes, or as needed so that it stays cold. Frozen meat, fish, or poultry can also be thawed in a microwave oven, if they will be cooked immediately after thawing.

If meat, fish, or poultry is cooked without thawing, additional time must be allowed. How much depends on the size and shape of the product. Large frozen roasts could take up to 1½ times as long.

When frozen meat, fish, or poultry are to be breaded and fried, they should be at least partially thawed in the refrigerator first for easier handling. All poultry that is to be stuffed should be thawed completely for safety. For best quality, cook thawed meat or fish immediately.

What to Do with Thawed Food after a Power Outage

If you lose power, keep the freezer door closed. Some thawed foods can be refrozen. However, the texture will not be as good. Here are some guidelines:

Meat and Poultry: Refreeze if the freezer temperature stays 40°F or below and if color and odor are good. Check each package, and discard any if you find signs of spoilage, such as an odd color or off odor. Discard any packages that are above 40°F or at room temperature.

Vegetables: Refreeze only if ice crystals are still present or if the freezer temperature is 40°F or below.

Discard any packages that show signs of spoilage or that have reached room temperature.

Fruits: Refreeze if they show no signs of spoilage. Thawed fruits may be used in cooking or making jellies, jams, or preserves. Fruits survive thawing with the least damage to quality.

Shellfish and Cooked Foods: Refreeze only if ice crystals are still present or the freezer is 40°F or below. If the temperature is above 40°F, throw these foods out.

Freezer Management

A full freezer is most energy efficient, and refilling your freezer several times a year is most cost efficient. If the freezer is filled and emptied only once each year, the energy cost per package is very high. You can lower the cost for each pound of stored food by filling and emptying your freezer two, three, and even more times each year.

Post a frozen-foods inventory near the freezer, and keep it up to date by listing the foods and dates of freezing as you put them in the freezer. Check them off as you take them out. By keeping an inventory, you will know the exact amounts and kinds of foods in the freezer at all times. It also helps to keep foods from being forgotten.

Organize the food in the freezer into food groups for ease in locating. Arrange packages so that those that have been in the freezer the longest are the first ones used.

Maintain the storage temperature at 0°F or lower. At higher temperatures, foods lose quality much faster. Keep a freezer thermometer in your freezer, and check the temperature frequently.

Smoking Fish

This delicious way of preserving a catch of fish is not without its pitfalls. Fish smoked without proper salting and cooking can cause food-borne illness—it can even be lethal. Many dangerous bacteria can and will grow under the conditions normally found in the preparation and storage of smoked fish. *Clostridium botulinum* (the cause of botulism) is the most notorious of these bacteria, but there are other harmful ones as well.

Because it is not easy for the average person to determine the final salt content of fish, the following parameters for adequate cooking while the fish is being smoked and refrigeration after the fish is smoked are the only ways a backyard homesteader can ensure a product will not support the growth of harmful bacteria:

- **You must heat the fish until the internal temperature reaches 150°F (preferably 160°F) and is maintained at this temperature for at least 30 minutes.**
- **You must salt or brine fish long enough to ensure that adequate salt is present throughout the smoked fish (at least 3.5 percent water-phase salt, the amount of salt based on the water in the fish).**
- **If storing, you must keep smoked fish under refrigeration at 38°F or less.**

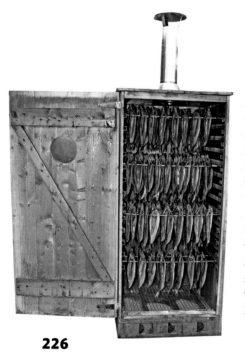

A smoker can be bought ready-made, or you can build one. Make sure it has all the features of a safe smoker, including controllable venting, a heat source for smoking, and a separate heat source for raising the temperature to 225°F.

The Right Wood

Use only hardwood for making smoke. Maple, oak, alder, hickory, birch, and fruit woods are all suitable for smoking fish. Wood from conifers leaves an unpleasant taste on the fish. Do not use fir, spruce, pine, or cedar.

Higher-fat fish absorb smoke faster and have better texture after smoking than lower-fat fish. Some of the ideal species for smoking are shad, sturgeon, smelt, herring, steelhead, salmon, mackerel, sablefish, and tuna.

You can smoke any fish without worrying about food-borne illness if you observe the basic principles explained below for preparation, salting, smoking, cooking, and storage.

Preparation

Different species of fish require different preparation techniques. Salmon are usually prepared by removing the backbone and splitting. Bottom fish are filleted. Small fish such as herring and smelt should be headed and gutted before brining.

Certain principles apply in all cases. First, use good-quality fish. Smoking will not improve fish quality; in fact, it may cover up certain conditions that could create food-safety problems later.

Thaw frozen fish in the refrigerator or in clean fresh water. Clean all fish thoroughly to remove blood, slime, and harmful bacteria. Keep fish as cool as possible at all times, but do not refreeze. When you cut fish for smoking, remember that pieces of uniform size and thickness will absorb salt in a similar way, reducing the chance that some pieces of fish will be either under- or over-salted. Do not let fish sit longer than 2 hours at room temperature after cleaning and before smoking.

Smoking and Cooking

Hot-smoked fish requires smoking followed by cooking. The length of smoking time depends on the flavor and moisture level you want. Smoking first will result in a better-tasting product due to less of a baked fish flavor and curd formation caused by juices boiling out of the fish. Smoke your fish for up to 2 hours at around 90°F in a smoker, and then increase the heat until the fish reaches a temperature of at least 150°F (preferably 160°F), and cook for at least 30 minutes. *It is important to measure fish temperature because of variations in how warm air circulates inside smokers.* A long-stemmed thermometer inserted into the thickest piece of fish through a hole in the smoker wall will allow temperature monitoring without opening the door. If the air temperature in your smoker cannot reach 200°F–225°F, you'll need to cook the fish in your kitchen oven within 2 hours after the smoking process. Waiting longer presents a danger of spoilage from bacterial growth. As in a smoker, the core temperature in the thickest piece of the fish must be maintained at a minimum 150°F for 30 minutes.

Remember: *Smoke itself is not an effective preservative under most conditions.*

When purchasing or building a smoker, here are the necessary features:

- **Independent source of heat for the pot of wood chips or logs**
- **Controllable vent or flue at the top**
- **Controllable draft at the bottom, thermostatic control over the oven temperature**
- **Another heat source to raise the temperature in the smoker to 225°F**

It is difficult to reach temperatures high enough for proper cooking with small metal smokers readily available in most hardware or sporting-goods stores. Supplement their use with oven cooking to achieve a core temperature for the fish of 150°F–160°F. A small metal smoker could be used for up to 3 hours to complete the first portion of the process, and then a home oven can safely complete the procedure by heating the fish to an internal temperature of 160°F for 30 minutes.

Refrigerating and Freezing Smoked Fish

Vacuum packed	Freeze or refrigerate (38°F or below)
Not vacuum packed	Refrigerate or freeze
Storage longer than 2 weeks	Freeze
Refrigerator temperature above 40°F	Freeze

Storage

Freeze or refrigerate (preferably at 38°F or less) your smoked fish if you vacuum pack it and do not plan to eat it immediately. This is essential if you have any doubt about the salt content or time and temperature process achieved during smoking. If you do not vacuum pack your smoked fish, it is important to keep it refrigerated to maintain both safety and the best product quality.

If storing longer than two weeks, tightly wrap and freeze smoked fish. Properly frozen, fish can hold for up to one year without any loss in flavor or quality.

Although regulations for commercial fish smokers may permit a minimum internal fish temperature lower than 160°F for the 30 minutes of cooking, home smokers don't have the continuous time and temperature recording equipment necessary to ensure proper cooking. Therefore, it is important to maintain these standards.

Similarly, if your home refrigerator cannot reliably maintain a temperature under 40°F, you should keep the smoked fish frozen regardless of whether or not it is vacuum packaged. Besides *Clostridium botulinum*, there are other dangerous bacteria that could potentially grow on smoked fish held in a refrigerator that cannot hold a cold enough temperature for prolong periods of time.

Smoking Meat

Where there's smoke, there's well-flavored meat and poultry. Using a smoker is one method of imparting natural smoke flavor to large cuts of meat, whole poultry, and turkey breasts. Smoking involves slowly cooking food indirectly over a fire.

Thaw Meat before Smoking

Completely thaw meat or poultry before smoking. Because smoking uses low temperatures to cook food, the meat will take too long to thaw in the smoker. (See "Safe Thawing," page 223.)

Using a Smoker

Cook food in appliances made for smoking. Don't smoke foods in makeshift containers such as galvanized steel cans or other materials not intended for cooking. Chemical residue contamination can result.

When using a charcoal-fired smoker, buy commercial charcoal briquettes or aromatic wood chips. Set the smoker in a well-lit, well-ventilated area away from trees, shrubbery, and buildings. Let the charcoal get red hot with gray ash—about 10 to 20 minutes depending on the quantity. Pile the charcoal around a drip pan, and fill the pan with water to maintain a moist environment. The drip pan catches any fat or juice from the meat or poultry and prevents it from flaming up on the coals. Add about 15 briquettes every hour, and check to make sure that the drip pan contains water. The most satisfactory smoke flavor is obtained by using hickory, apple, or maple wood chips or flakes. Soak the chips in water to prevent flare-ups, and add about ½ cup of chips to the charcoal as desired.

Using a Covered Grill

To smoke meat and poultry in a covered grill, pile about 50 briquettes in the center of the heat grate. When they are covered with gray ash, push them into two piles, and center a drip pan filled with water between them. Close the lid, and keep the grill vents open. Add about 10 briquettes every hour to maintain the temperature.

Chill Promptly

Refrigerate meat and poultry within two hours of removing it from a smoker. Cut the meat or poultry into smaller portions or slices; place it in shallow containers; and refrigerate it. Use it within four days, or freeze it.

Use Two Thermometers to Smoke Food Safely

To ensure that meat and poultry are smoked safely, you'll need two types of thermometers: one for the food and one for the smoker to monitor the air temperature. The smoker's thermometer should reach a maximum of 225°F—the high temperature on most commercial smokers. (The USDA recommends a high temperature of 300°F.) Many smokers have built-in thermometers, but add-on units are available.

Use a food thermometer to determine the temperature of the meat or poultry. You can insert an oven-safe thermometer in the meat and let it remain there during smoking or use an instant-read thermometer.

Cooking time depends on many factors: the type of meat, its size and shape, the distance of the food from the heat, the temperature of the coals, and the weather. It can take anywhere from 4 to 8 hours to smoke meat or poultry, so it's imperative to use thermometers to monitor temperatures.

Smoke food to a safe minimal internal temperature:

- Beef, veal, and lamb steaks, roasts, and chops may be cooked to 145°F.
- All cuts of pork should be heated to 160°F.
- Ground beef, veal, and lamb may be cooked to 160°F.
- All poultry should reach a safe minimum internal temperature of 165°F.

Brewing Beer

An estimated 750,000 Americans enjoy home brewing, a testament to the joys of crafting your own tipple. For the backyard homesteader, home brewing is yet another step toward independence and self-sufficiency—not to mention pride in producing a pure beverage that may well measure up to some of the finest microbrews. Fogging up the kitchen windows on a winter's night with the rich flavors of malted grain is a treat in itself.

The allied skills of wine and cider making offer an economic edge if you use your own produce, whether grapes, flowers, or apples. It's pretty sweet to have your grape arbor deliver a delicious beverage to your table. These pursuits are fun to do and loaded with anticipation as you wait for your brew to mature into perfection.

Be prepared to invest in some special equipment that will set you back $80 or more. And because the ingredients will likely be bought, a case of your own brew will cost between $12 and $23 each—a savings only if you favor boutique brews, which of course is what you are producing. Finally, although the brewing steps are quite simple and even fun, you'll be doing a lot washing—pans, siphons, "carboys" (think water-cooler jugs), and bottles, bottles, bottles.

Fermentation is fully underway in this carboy of dark beer. The air lock releases CO_2 without allowing contaminated air in.

Making a Batch

A basic batch of beer is 5 gallons, enough to fill about 48 12-ounce bottles—two cases' worth. Home-brew supply shops and online sources offer basic equipment kits that include the following:

- 6.5-gallon fermenting bucket with lid
- 6.5-gallon bottling bucket with spigot
- Air lock
- Siphon
- Bottle filler
- Hydrometer
- Twin-lever capper
- Bottle brush
- Bottle caps
- Instructions, perhaps a DVD

In addition, you'll need a 3–5 gallon stockpot and a cooking thermometer. Supplies include malt extract syrup, crushed grains, hops, sugar, and yeast.

The actual brewing begins by dropping a muslin bag of crushed grain into about a gallon of water and bringing it to 160°F, holding the temperature for a prescribed amount of time, and then removing the grain. Next, add the malt-extract syrup—wonderful stuff—and stir until it dissolves completely. Boil; then add the hops; and boil some more.

You now have something called wort, the stuff of which beer is made. It, along with cold water, goes into the fermentation carboy (which, of course, you have washed thoroughly in advance). At this point you'll take a hydrometer reading, the first of several, and make sure it is in accord with the recipe you're following. Next, sprinkle in the yeast and attach the fermentation air lock, a water-filled gizmo that lets CO_2 out but won't let in contaminating air. Within a day the wort will start bubbling and continue working for about 7 days.

Bottling begins with boiling sugar in water: the stuff the yeast will feed on to produce carbonation and alcohol. Pour the sugar water into the bottling bucket; filter and siphon the beer in. Fill bottles and cap them. Mop up the mess you've made on the floor.

Store the beer at 67°F–70°F for at least 10 days before popping a bottle open and savoring your brew.

Making Wine

Even if you buy all the ingredients, including the raw grape juice (known in the trade as "must"), you can make your own wine for $2 a bottle. If you use your own fruit to make the must, equipment and a few preservatives and clarifiers are your only expense.

As with beer, buying a starter kit of equipment is an affordable way to equip yourself for wine making. For around $100, a starter kit includes:

If you have many grape vines you may want to invest in a wine crusher. A plastic stock-watering tank catches the crushed grapes.

- 6-gallon glass carboy
- 8-gallon fermenter with lid
- Double-lever corker
- Bottle filler
- Thermometer
- Air lock
- Sterilizer
- Hydrometer
- Siphon
- Corks
- Instructions or DVD

However, many wine makers do well with much less, relying on a glass carboy for fermenting and using a quart measuring cup for filling bottles. To decide the approach that is right for you, visit a home-brewing shop (which seem to be everywhere), check some online videos, or talk to an experienced home brewer.

Making a Batch

The raw material for wine is crushed fruit or berries, prepared in one of four ways:

Cold maceration involves a soup of the fruit, sugar, water, and sulfites. A pectin enzyme is added, and the whole brew is refrigerated for up to two days before yeast is added.

Hot water extraction starts with pouring the crushed grapes into the fermentation jug before sugar and hot water are added. After stirring, the mix is cooled; then yeast is added.

With **direct-heat extraction,** the crushed fruit and a small amount of water are cooked in a pan. Sometimes the pulp is strained out; sometimes it is left in before sugar and water are added as prep for primary fermentation.

Fermentation extraction is the simplest process. The crushed fruit is poured into the fermentation jug; then water and any other ingredients are allowed to work before the yeast is added.

The product of each of these processes is a rich soup called *must*. One very simple approach is to mix one-third crushed pulp with one-third sugar, blended in one-third water. Wine recipes vary widely. Some combine fruits like strawberries, blueberries, and grapes. Others add things like oak chips (key to Chardonnay) and other flavoring ingredients. Part of your journey of exploration will be discovering the recipes that best suit your produce and your palette. All follow this basic pattern:

First, wash the fruit and remove all stems and leaves. Next, crush the fruit. Using the bottom of a drinking glass is a simple method, preferable to using one's hands or feet. Or if you have a lot of grapes, you can buy a mechanical crusher. From this pulp, make the must using one of the methods described above.

Wash all the tools involved in the process using a sanitizer intended for wine making. (Soap and bleach can kill yeast and add things you don't want to your wine.)

Add this to the primary fermenter (whether a carboy or plastic bin), and add water. Temperatures are important, so follow the recipe carefully. A 6-gallon fermenter will yield 28 bottles of wine. Often Bentonite, a clay-like clarifying agent, is added at this early stage. Many recipes have you use a hydrometer to check the amount of sugar and therefore the anticipated alcohol level, something you may be asked to do at several stages. Now is the time you sprinkle in yeast and install the air lock. Within two days, fermentation starts and continues for up to a week.

Next, transfer the wine to another container, leaving any sludge or additives behind, taking care to expose the wine to as little air as possible as you do so. Install the airlock, and give this second fermentation a week to six weeks to work, depending on your recipe.

Finally, transfer the wine to leave any remaining sludge behind. Some wine makers filter the wine through a clean cotton cloth. Before bottling you might add sulfite, a preservative that also stops fermentation, as well as isinglass, another clarifier. From this container, bottle and cork the wine. An inexpensive wand siphon is worth the modest investment. A double-lever corker costs as little as $30; push in types run as little as $10. Corks in lots of 100 cost about $15. Now, cellar your wine and wait for the magic to happen.

Making Apple Cider

In the United States, fermented alcoholic apple juice is called "hard cider," while freshly pressed, nonalcoholic cider is called "sweet cider." Cider is made from fermenting apple juice, which relies on natural yeast present in the apples for fermenting.

Fresh or unpasteurized apple juice or cider can cause food borne illness from bacteria. The directions that follow will help you make and store apple cider safely.

Selection of Apples

Apples used for cider don't have to be flawless. They do, however, have to be free from spoilage. Spoiled areas will cause the juice to ferment too rapidly and will ruin the cider. Don't use apples that appear brown, decayed, or moldy, but blemished apples and small-sized apples can be used. You can mix apple varieties together or use all one variety.

Apples should be firm and ripe. Green, under-mature apples cause a flat flavor when juiced. The best cider comes from a blend of sweet, tart, and aromatic apple varieties. A bushel of apples yields about 3 gallons of juice.

Wash glass jars or bottles in warm, soapy water. Rinse thoroughly so that no soap remains. Prepare a clean muslin sack or jelly bag for juicing the apples. If you're using a new muslin sack or pillowcase, wash it first to remove any sizing. Be sure all soap is rinsed out. An old but absolutely clean pillowcase will work.

Always, as with any food preparation, start by washing your hands and forearms thoroughly with hot water and soap for at least 20 to 30 seconds. Utensils and equipment can be easily sanitized after washing and rinsing by filling with or soaking in a mixture of 1 tablespoon household bleach per gallon of warm water for at least 1 minute.

Juicing Apples

Small household appliances can be used to juice your apples. You'll need to core and cut the apples, then process them through a food chopper, blender, or food processor. Put the crushed apple pulp into a clean muslin sack or jelly bag, and squeeze out the juice.

If you want to drink the juice now without making cider, pasteurize it by heating it to at least 160°F. Then pour juice into clean glass jars or bottles, and refrigerate it.

Pasteurizing and Storing Cider

Unpasteurized, or fresh, cider may contain bacteria, such as *E. coli O157:H7* or Salmonella, that can cause illness. Harmful bacteria must be killed by a pasteurization process prior to drinking the cider. To pasteurize, heat cider to at least 160°F, 185°F at most. Measure the actual temperature using a cooking thermometer. It will taste less "cooked" if it is not boiled. Skim off the foam that may have developed, and pour the hot cider into heated, cleaned, and sanitized plastic containers or glass jars. Refrigerate immediately.

To freeze it, pour hot cider into plastic or glass freezer containers, leaving a ½-inch headspace for expansion. Refrigerate the cider until cool, and then place it in the freezer.

Making Sweet Cider

Begin with freshly pressed juice (not pasteurized). If clear cider is what you want, let the bottled cider stand at 72°F for 3 to 4 days.

Clean bottles should be filled to just below the brim and stoppered with new, clean cotton plugs instead of a regular lid or cap. The cotton plug is used for safety. If pressure builds up during the fermentation that occurs, the cotton will pop out and release the pressure. If you place a cap on the bottle, the bottle may explode.

After 3 or 4 days, sediment will begin settling on the bottom as fermentation bubbles rise to the top. Now is the time to stop fermentation if you want sweet, mild cider. Extract the clear liquid from the sediment by "racking off" the cider, transferring it to a clean container while leaving the sludge behind, much as you would beer or wine.

Pasteurize the cider to ensure its safety by heating to at least 160°F. Store the cider in the refrigerator at 40°F or lower, and drink it within 5 days. Freeze the cider after pasteurization for longer storage.

For larger quantities of apples, consider using a fruit press, opposite. Follow the directions that accompany your press for juicing apples.

Before juicing apples, sort and wash them well under clean running water. Discard spoiled apples. Core and cut the apples into quarters or smaller pieces, above.

Fermenting Dry, or Hard, Cider

If you are seeking a dry cider, ferment the cider longer at room temperature. Use an air lock on a carboy. In about 10 days, the cider will begin looking frothy and may foam over the top. This is normal; simply wipe off the bottle; replace the cotton with clean fabric; and let frothing continue until the foaming stops, which signals that fermentation is complete.

Tips for Safe Cider

- Avoid using apples that have visible signs of decay or mold growth.
- Wash apples thoroughly before pressing or grinding to make cider.
- Use a fruit press or small kitchen appliances to crush the fruit.
- Start by washing your hands and sanitizing equipment. Place washed and rinsed utensils and equipment in a mixture of 1 tablespoon household bleach per gallon of water for at least 1 minute.
- Squeeze juice through a clean, damp muslin cloth.
- Pasteurize cider to ensure safety. Heat to at least 160°F, and pour into warm jars to prevent breakage.
- Store cider in the refrigerator for immediate use, or up to 5 days. Put cider in the freezer if you want to keep it longer.
- Use strong, sound glass bottles that will not break during fermentation.

Caution: Young children, elderly, and immunocompromised individuals should never drink fresh apple cider unless it has been heated to at least 160°F.

Root Cellars

Preserving food by burying it in the ground has long been a preferred way to store the harvest. The method works because the temperature belowground is relatively stable and is always cooler than summer temperatures. Just as important, it protects produce from freezing in the winter.

Underground temperatures vary considerably across the U.S., so assess your situation to see whether a cellar will work in your location. For example, while Maine might have a mean earth temperature (the temperature of the earth 30 feet down) of 46°F, the mean temperature is 67°F in central California—too high a temperature to hold perishable fruit for long. However, a root cellar is likely to serve a useful purpose no matter how warm the temperature. For example, you'll have to dig mighty deep to strike cool temperatures in Arizona, but a root cellar there can be ideal for storing nuts and dried fruit. Before

A great idea whose time has come around again, a root cellar uses the moderating effect of ground temperature to keep produce cool in warm seasons and keep it from freezing during the winter.

breaking ground, check with your County Extension Office if you have doubts about your locale's suitability.

A root cellar can be as simple as a plastic garbage bin buried in the ground and covered by a couple of straw

Mini Root Cellar

Simplicity itself, a plastic trash can set in a hole can store a modest amount of produce through the winter. Setting a couple of straw bales over it insulates it from wintry blasts. To avoid having to lie in the snow to reach the last of your bounty, place the produce in net bags attached to a length of rope.

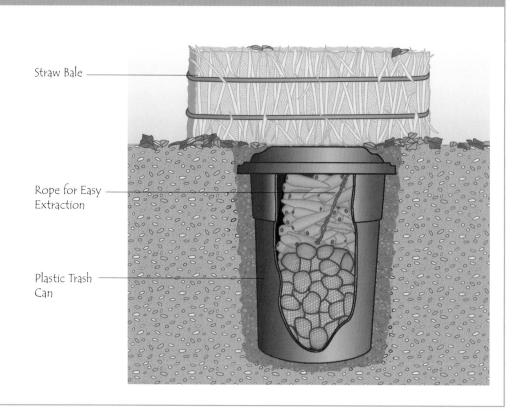

Straw Bale

Rope for Easy Extraction

Plastic Trash Can

bales or as elaborate as a buried 40-foot-long shipping container. And surprising things make suitable root cellars. One innovator buried a school bus, another sunk an old fiberglass sailboat hull in his backyard. And here's a sweet reversal: remove the cooling system from a refrigerator, and bury the cabinet door up. Once installed, a cellar costs next to nothing to maintain and can safely store your harvest through the winter.

Ideal Produce for the Root Cellar

The farm cellar of yore stored everything—apples, potatoes, canned produce, cider, and even cured hams. The results were sometimes mixed. Ladling off the rotten upper layer of a crock of sauerkraut or making the best of shriveled potatoes was something you just put up with for lack of any better storage options. Today, we can pick and choose what we store in a root cellar. Here are some ideal produce candidates, what farmers used to call good "keepers":

- **Root crops like potatoes, beets, turnips, kohlrabi, carrots**
- **Apples, plums, peaches, though they emit ethylene gas, which affects potatoes and carrots**
- **Winter squash, pumpkins**
- **Beans**
- **Onions, garlic**
- **Nuts**

With careful handling and in the right climate, even perishables like cheese and butter can be stored in a root cellar.

Air Circulation

For any root cellar larger than a trash can buried in the ground, install some means for circulating the air. With at least two vents, one to pull in air, one to exhaust it, you can even out temperature and control humidity. In addition, you can get rid of some of the ethylene gas and odors that may affect your produce. Make sure any vent openings are covered with hardware cloth to keep out rodents. Painting a vent black creates some convection (from solar heat) and boosts the temperature in the depth of winter. Baffles on the vents let you adjust for seasonal variations. Install a thermometer as a guide. Bear in mind that warm air rises; the ceiling of your root cellar might be 10°F warmer than the floor. Circulation can even out that difference. Good air circulation also helps control humidity, which is a blessing and a bane for root cellars. (See below.)

Controlling Humidity

A hole in the ground is inevitably humid. That's generally a good thing because it keeps fruit and vegetables from drying out. However, too much humidity can be a problem for your root cellar. What you don't want is the ceiling sweating moisture on your carefully harvested produce. And if you hope to store canning jars full of produce, you'll want limited humidity so that the lids don't rust, compromising the vacuum seals.

To lower humidity, here are your options:
- Add a container of rock salt to wick up the moisture.
- Replace an earthen floor with concrete.
- Make sure there is adequate drainage around the cellar.
- Ramp up air circulation.
- Lessen the effects of humidity by installing shelves away from the walls and leaving enough space between items for air to circulate.

To raise humidity:
- Sprinkle water on the floor.
- Reduce air circulation.
- Pack produce in wet sand or sawdust.

Installing a hygrometer, which indicates the humidity level, will tell you if your tinkering is successful.

Root-Cellar Construction

An architect once disparagingly referred to building a basement as digging a well and then trying to keep the water out. That's not too far off the mark for a root cellar. The whole purpose of a root cellar is to take advantage of the natural temperature moderation of the earth. Exploit it to the fullest. If a hillside is handy, you can burrow into it and still have convenient access with some stairs and a doorway. If your property is flat, you'll have to dig deeper and install a long stairwell to reach the cellar. In either case, locate the door away from the sun.

Anything with structural integrity can be used as the core of a root cellar. Steel culvert piping is relatively portable and made to be buried. Concrete culvert is a fine choice as well, but it is expensive to haul in and position. Formed concrete and concrete blocks on a concrete footing are ideal building materials if they are within your budget. Because you'll likely want to cover your root cellar with earth—even 1 foot of soil can lower the temperature up to 20°F—you'll need a substantial ceiling of reinforced concrete or pressure-treated plywood supported by laminated beams.

Allow for drainage by adding perforated drainpipe around the top of the cellar. In extreme conditions you may need to add drainpipe around the perimeter of the floor that flows into a sump well equipped with a pump. This will also control hydrostatic pressure—the considerable force of water-saturated soil against the walls of the cellar.

Add insulation to your walls to enhance the insulating capacity of the soil. High-density foam insulation panels work best because they are unaffected by moisture and can be applied on the inside or outside of the walls. If possible, leave an air space between the walls of your cellar and the surrounding earth to reduce pressure on the walls and add further insulating effect. However, be sure the roof covers the gap on all sides so that rainwater won't erode the sides of the excavation.

Basement Cool Room

If you have a basement, consider saving yourself a lot of digging by turning a corner of it into a cool room, right—a close cousin of the root cellar. A basement is not cold enough to be a root cellar, but by letting some winter air into an insulated room you can have the convenience of storing produce indoors without the cost of refrigeration. A key ingredient is a handy window that lets you run two ducts outside. Removing the window frame and adding a plywood insert does the job. Equip each duct with a baffle so that you can adjust the exchange of air. Run one duct nearly to floor level for incoming cold air. Insulate the walls, floor, and ceiling of the room. An inexpensive hygrometer/thermometer lets you monitor humidity and temperature.

Basement Cool Room

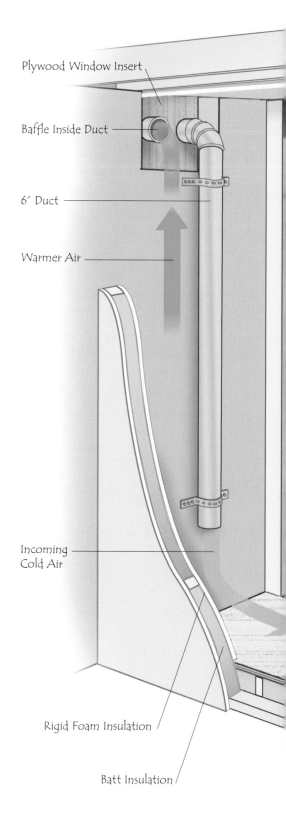

Plywood Window Insert

Baffle Inside Duct

6" Duct

Warmer Air

Incoming Cold Air

Rigid Foam Insulation

Batt Insulation

Hygrometer/
Thermometer

Resources

American Dairy Goat Association

P. O. Box 865
161 W. Main St.
Spindale, NC 28160
828-286-3801
www.adga.org
Though aimed at commercial producers, this site is loaded with useful information about raising goats for milk. The link "About Dairy Goats" is a great introduction to raising milk-type goats.

American Homebrewers Association

P.O. Box 1679
Boulder, CO 80306
888-822-6273
www.homebrewersassociaton.com
This "community" for homebrewers offers information, events, and competitions.

American Livestock Breeds Conservancy

P. O. Box 477
Pittsboro, NC 27312
919-542-5704
http://albc-usa.org
A nonprofit organization, ALBC works to protect over 180 breeds of livestock and poultry from extinction. Included are asses, cattle, goats, horses, sheep, pigs, rabbits, chickens, ducks, geese, and turkeys.

BackYardChickens.com

www.backyardchickens.com
This website is loaded with ideas and information, with valuable advice and photos from other chicken raisers. Read "Raising Chickens 101" for starter information on why and how to raise chickens.

BackYardHive.com

P. O. Box 333
Eldorado Springs, CO 80025
888-297-5374
www.backyardhive.com
You'll find plenty of general beekeeping info at this site, with an emphasis on the Top-Bar hive.

BioWorks, Inc.

100 Rawson Rd. Ste. 205
Victor, NY 14564
800-877-9443
www.bioworksinc.com
Bioworks is a manufacturer of environmentally responsible pesticides.

Burpee

W. Atlee Burpee & Co.
300 Park Ave.
Warminster, PA 18974
800-333-5808
www.burpee.com
This well-known company offers a planting seed catalog and a wide variety of seeds from which to choose.

ChickenCoopSource.com

Hayneedle, Inc.
9394 W. Dodge Rd., Ste 300
Omaha, NE 68114-3319
866-749-9504
www.chickencoopsource.com
Prefer to buy a chicken coop as a kit or ready-made? Here's a great source for coops, tractors, and all things chicken—plus helpful background on raising chickens.

El Dorado Heirloom Seeds

Lipscomb Enterprises
206 E Central Ave.
El Dorado, KS 67042
316-452-5582
http://eldoradoheirloomseeds.com
This company offers non-hybrid, open pollinated, organic heirloom seeds.

Feeding America

35 E. Wacker Dr., Ste. 2000
Chicago, IL 60601
800-771-2303
www.feedingamerica.org
This organization is the largest domestic hunger-relief charity in the U.S.

Department of Horticulture

Cornell University
134A Plant Sciences Building
Ithaca, NY 14853
607-255-4568/1789
www.gardening.cornell.edu
This site includes growing guides for more than 50 vegetables, a garden tour, and fact sheets on basic techniques, insect pests, plant diseases, and common problems.

Fruit Production for the Home Gardener

Penn State College of Agricultural Sciences
201 Agricultural Administration Building
University Park, PA 16802
814-865-2541
http://agsci.psu.edu/fphg
Besides a thorough introduction to growing fruit, from planting to harvesting, this site includes sources of nursery stock.

Gardens Alive!

5100 Schenley Pl.
Lawrenceberg, IN 47025
513-354-1482
www.gardensalive.com
This online retailer sells environmentally-responsible gardening products.

GreenGrid

Weston Solutions, Inc.
750 E. Bunker Ct., Ste. 500
Vernon Hills, IL 60061
847-918-4000
www.greengridroofs.com
GreenGrid provides a customized modular system for turning roofs into roof gardens.

HobbyFarms.com

P.O. Box 8237
Lexington, KY 40533
hobbyfarms@bowtieinc.com
800-627-6157
www.hobbyfarms.com
From Hobby Farms magazine, this entertaining and informative site offers a range of topics relevant to an aspiring backyard farmer. The livestock breed profiles are a handy start in choosing what animals might be right for you.

IrrigationTutorials.com

www.irrigationtutorials.com
These detailed tutorials on planning and installing drip-irrigation systems include reviews of various product types.

The Modern Homestead

www.themodernhomestead.us/article/Butchering-Ready.html
Next to watching it done, this site, featuring clear photos and straightforward commentary, may be your best guide to butchering chickens.

Resources

My Pet Chicken

501 Westport Ave #311

Norwalk, CT 06851

888-460-1529

www.mypetchicken.com

This entertaining site is aimed at keepers of backyard chickens, with plenty of information and access to everything you need to get started, including baby chicks, chicken coops, accessories, books, and gifts.

NCAT Sustainable Agriculture Project

P.O. Box 3838

Butte, MT 59702

https://attra.ncat.org

Focusing on sustainable and organic production methods for traditional produce, this site also introduces alternative crops and enterprises.

National Center for Home Food Preservation

University of Georgia

208 Hoke Smith Annex

Athens, GA 30602-4356

Fax: 706-542-1979

www.uga.edu/nchfp

This is the authoritative source for research-based recommendations for home-food preservation. The Center was established with funding from the Cooperative State Research, Education and Extension Service, U.S. Department of Agriculture (CSREES-USDA).

National Gardening Center

1100 Dorset St.

South Burlington, VT 05403

802-863-5251

www.garden.org

In addition to a world of attractively presented information, this site offers helpful regional updates as the seasons change and information on seed swaps.

National Honey Board

11409 Business Park Cle.

Ste. 210

Firestone, CO 80504-9200

303-776-2337

www.honey.com

Need some honey recipes or background on the many uses of honey? This attractive site offers a great education in its sweet virtues.

National Pygmy Goat Association

1932 149th Ave. SE

Snohomish, WA 98290

425-334-6506

www.npga-pygmy.com

While many goat breeds are worth considering for a backyard flock, pygmy goats best suit small spaces. This site offers resources of health, husbandry, and breeder contacts.

Omlet

Tuthill Park

Wardington, Oxfordshire, OX17 1RR

Great Britain

1-86-OMLET-USA

www.omlet.us

Omlet manufactures a stylish range of chicken houses, rabbit houses, and beehives. The site also offers helpful information on breeds and raising techniques.

Organic Honey Standards

Quality Assurance International

12526 High Bluff Dr. Ste. 300

San Diego, CA 92130

858-792-3531

www.apiservices.com/articles/us/organic_standards.htm

Though aimed at commercial honey producers, this site offers backyard beekeepers guidance on what organic honey entails.

Park Seed Co.

1 Parkton Ave.

Greenwood, SC 29647

800-845-3369

www.parkseed.com

Park is a long-established and well-known company in the mail-order seed and plant business.

PoultryU

University of Minnesota

1364 Eckles Ave.

305 Haecker Hall

St. Paul, MN 55108-6118

612-624-2722

www.ansci.umn.edu/poultry/resources/diseases.htm

This site will provide more than you want to know about the diseases chickens can contract. Articles pulled from a variety of sources will help you spot diseases and, when you call in the veterinarian, better understand the specified treatment.

Rearing Chicks and Pullets for the Small Laying Flock

University of Minnesota Extension

Office of the Dean

240 Coffey Hall

1420 Eckles Ave.

St. Paul, MN 55108-6068

612-624-1222

www.extension.umn.edu/distribution/livestocksystems/DI1191.html

You'll find clear and precise information about how to raise chicks for a small egg-laying flock at this site.

Soil and Plant Tissue Testing Laboratory

West Experimentation Station

682 N. Pleasant St.

University of Massachusetts

Amherst, MA 01003

413-545-2311

www.umass.edu/soiltest

The lab will conduct a variety of soil or plant tests on samples it recieves.

Southside Farms

522 Hall Rd.

Worcester, NY 12197

518-588-0188

www.southsidestables.com

This site is a good introduction to miniature milk cows.

Stalite PermaTill

877-737-6284

www.permatill.com

Stalite manufactures a slate-based lightweight expanded aggregate to amend compacted soils.

WinePress.US

www.winepress.us

This handy grape-growing and wine-making forum is loaded with information.

Glossary

Acid food Fruits and vegetables that have a 4.6 pH or lower. Acid foods may be processed in a boiling-water canner and don't require pressure canning.

Aerobic fermentation The fermentation process in which a crock, vat, or pail is in contact with air so that the yeast can reproduce rapidly.

Air lock A device used to protect fermentation from contaminating air. It allows carbon dioxide to exhaust while keeping air out.

Annual A plant that germinates, grows, flowers, produces seeds, and dies in the course of a single growing season; a plant that is treated like an annual and grown for a single season's display.

Ascorbic acid The chemical name for vitamin C. When added to foods before canning, ascorbic acid preserves natural color.

Banding The application of tight latex rings to livestock for the purpose of castration, tail docking, or the removal of horn buds.

Bantam A breed of small chicken whose mature size can be less than half that of a typical chicken.

Barley A grain used to make the mash that is the foundation for brewing beer.

Bearding A congregation of bees at the front of the hive.

Bee brush A soft brush used to remove bees from combs while harvesting honey.

Blanch Scalding to loosen skins and preserve the color of fruits and vegetables before drying.

Blind teat A teat that will not yield milk due to disease, injury, or birth defect.

Bloom The protective coating on an egg that keeps bacteria from penetrating the shell.

Blowout A serious vent injury to a chicken caused by laying an over-large egg.

Boiling-water canner A lidded kettle with a jar rack used for processing jars of food in boiling water.

Bottling wand A plastic tube with a pushdown valve at one end and a length of plastic tubing at the other end, used to fill bottlers with minimum mess and exposure to the air.

Bottom-fermenting yeast Also referred to as "lager yeast," this brewing yeast works at low temperatures and produces a clean, crisp taste. After fermentation, it settles to the bottom of the brewing container.

Botulism A potentially fatal illness caused by eating the toxin produced by the growth of *Clostridium botulinum*. The bacteria grows in moist, low-acid foods containing less than 2 percent oxygen and stored between 40° and 120° F. Proper heat processing destroys this bacterium in canned food.

Breeders Chickens raised expressly for the commercial production of fertilized eggs.

Broadcast seeding A planting technique that involves randomly scattering seeds over prepared soil rather than sowing in rows. Most often used with small plantings of crops such as carrots, beets, radishes, and salad greens.

Brood Bee eggs, larvae, or pupae still in their cells.

Brood chamber or nest The area of a hive where bees are reared.

Brooder A dry, heated enclosure designed to protect young chicks.

Broody A laying hen's desire to stay on the nest and incubate her eggs.

Bubbler An air lock made of plastic or glass that protects the brewing process from contaminating air while allowing the release of the carbon dioxide produced during fermentation.

Buck A male goat.

Buckling A young male goat.

Bung A stopper that seals a brewing keg or cask after it is filled.

Burr comb Undesirable comb built by bees in a place that obstructs the removal of frames.

Canning A method of preserving food in airtight containers. Heat kills bacteria and creates a vacuum to safely seal the container.

Canning salt Also called pickling salt, it does not have the anti-caking or iodine additives of table salt.

Capon A rooster that has been castrated to increase his weight gain and desirability as a meat source.

Capped honey Honey-filled comb cells covered by bees with a thin layer of wax.

Caprine An animal related to the goat or antelope family.

Carboy A small-mouthed glass or plastic container of 2 to 7 gallons used for brewing.

Clarify A stage in wine making when the yeast and tiny bits of pulp have settled to the bottom of the container. May include adding a powdered clay substance like Bentonite to hasten clarification.

Cloaca The single opening in a chicken for the bowel, urinary, and reproductive tracts.

Cloche A transparent or translucent cover used to protect plants from the cold.

Cockerel A juvenile male chicken less than one year old.

Cold packed A preserving method in which hot liquid is added to raw food in sterilized jars—also called raw packed.

Colony Honeybees and their queen, with or without males, living together. Also referred to as a nest or hive.

Colony Collapse Disorder (CCD) A syndrome of undetermined cause that strikes commercial bees trucked for the pollination of vegetable farms, killing whole hives of bees and threatening food production.

Colostrum The first milk a mammal produces, full of antibodies essential for their offspring.

Companion planting Positioning plants in the garden to take advantage of their influence on neighboring plants. Can be used to stimulate growth or to ward off pests or disease.

Compost A humus-rich, organic material formed by the decomposition of leaves, grass clippings, and other organic materials. Used to improve soil.

Creep An enclosure that small animals can enter for feeding but large animals cannot.

Cud Food regurgitated by a ruminant for rechewing.

Cull To remove livestock that are not productive or threaten the health of other animals.

Cultivar A cultivated variety of a plant, often bred or selected for some special trait, such as double flowers, compact growth, cold hardiness, or disease resistance.

Dam A mother goat.

Glossary

Debeak The process of blunting a chicken's beak through cauterization to prevent injury to other birds, typically used in commercial layer farms where dense housing is the norm.

Disbudding The removal of horn buds on a young goat to prevent the growth of horns. Cauterizing is the most common method.

Doe A female goat.

Doeling A young female goat.

Double dig To work soil to a depth that is twice the usual by digging a trench, loosening the soil at the bottom of the trench, then returning the top layer of soil to the trench. This produces a bed with a deep layer of loose, fluffy soil.

Drenching An oral administration of medicine to an animal.

Drone A male bee whose sole function is to mate with the queen. After the summer mating season, drones are starved and expelled from the hive.

Dust bath The use of dry dirt by a chicken to clean feathers and remove parasites.

Easement A category of municipal code that defends access to utilities and might include the right to share a driveway, parkway, or a public-access pathway running along a property.

Egg tooth The sharp point on a chick's beak that helps poke a hole in the eggshell during hatching.

Enterotoxemia Often called overeating disease, this malady in goats results from sudden exponential growth of otherwise beneficial bacteria. The bacteria produces a toxin that can lead to profuse diarrhea and discomfort, sometimes killing younger animals.

Estrus The period of time, usually 24 to 36 hours, when the female is sexually receptive to the male.

Fermentation The action of yeast upon sugar to produce alcohol and carbon dioxide.

Fermentation bottle A small-mouthed bottle equipped with an air lock used to complete fermentation. Also called a secondary fermentation vessel.

Fining Clarifying cloudy wine by blending clear wine, altering the temperature, filtering, or adding Bentonite, Isinglass, gelatin, or casein.

Flight path The predominant route used by bees as they travel to and from sources of nectar, ideally oriented away from human traffic.

Foot rot A condition in goats resulting from a fungus infection, best prevented by regular hoof trimming and access to dry pens.

Foulbrood A bacterial disease that attacks bee larvae.

Foundation Manufactured sheets of beeswax molded in the shape of a natural honeycomb, used as starters to guide bees in making uniform combs.

Frames Wooden structures configured according to the type of hive on which bees build their comb.

Free-range The method of raising chickens that affords them some access to a yard or pasture. The extent and duration of ranging can vary widely.

Freshen The process of a doe giving birth and producing milk.

Fruit leather A tasty, chewy, dried fruit product made by pouring pureed fruit onto a flat surface for drying.

GFCI (ground-fault circuit interrupter) A safety device that breaks a circuit when it senses a difference in flow between line and ground current. Building codes universally require such protection for outdoor electrical circuits.

GFCI circuit breaker A combination circuit breaker and GFCI installed in the service panel in place of a regular circuit breaker. It monitors current flow in both hot and neutral wires. When the breaker detects unequal current in the wires, it immediately shuts off power and protects the entire circuit.

Green honey Nectar or partially cured honey not evaporated to the extent that it can be capped.

Grit Ground granite or bits of sand stored in a chicken's crop to aid in breaking down food.

Hardiness zone Geographic region where the coldest temperature in an average winter falls within a certain range, such as between 0° and –10°F.

Hen A female chicken 1 year of age or greater.

Headspace The unfilled space above food or liquid in a jar that allows for expansion as the jar is heated and forms a vacuum as the jar cools.

Heat processing Treatment of jars with sufficient heat to enable storing food at normal home temperatures.

Hermetic seal An airtight seal that prevents reentry of air or microorganisms into canned goods.

Hive body The box that makes up the outer shell of the hive in which the frames are placed.

Hive stand A base made of wood or masonry that holds the hive off the ground, away from moisture and insects.

Honey bound The filling of the brood nest with honey in preparation for winter, sometimes a signal the colony is about to swarm.

Honey extractor A machine that uses centrifugal force to remove honey from the cells of a honeycomb.

Hops A vine-grown herb that is added to wort, or fermenting beer, to lend a pleasantly bitter aroma and taste.

Hot pack Foods that are cooked and hot when placed in a canning jar.

Humus The complex, organic residue of decayed plant matter in soil.

Hydrometer An instrument used to determine the potential alcohol content when brewing.

Hygrometer An instrument that indicates the humidity level in the air.

Incubator A container of controlled heat and humidity used to hatch eggs.

Isinglass A powdered substance used to clarify wine or beer.

Kenya Top-Bar hive A simple hive shaped like a trough in which combs hang vertically. It is covered by a lid-like roof and has legs to keep it at a convenient working height.

Langstroth hive Invented in 1851 by Lorenzo Langstroth, this is the standard hive used by beekeepers. Langstroth pioneered the use of add-on boxes to provide for honey production and the use of "bee space"— the ¼-to-⅜-inch space between honeycombs found in natural hives.

Let down A release of milk by a goat or sheep when nuzzled by young or stimulated by a milker.

Lot Coverage Ratio (LCR) The proportion of a residential property occupied by buildings and paving.

Glossary

Low-acid Foods that have a pH above 4.6. The acidity in such foods is insufficient to prevent the growth of the bacterium *Clostridium botulinum*. Vegetables, some tomatoes, figs, all meats, fish, seafood, and some dairy foods are low acid. To control the risk of botulism, jars of these foods must be heat processed in a pressure canner or acidified to a pH of 4.6 or lower before processing in boiling water.

Malt extract A sweet syrup or powder of condensed malted barley used in brewing.

Marek's disease A common poultry viral disease preventable by vaccinating chickens when they hatch.

Mastitis An inflammation of mammary glands, resulting in discomfort and reduced milk production.

Microclimate Local conditions of shade, exposure, wind, drainage, and other factors that affect plant growth at any particular site.

Milk fever An elevation of temperature in goats after giving birth, sometimes resulting in loss of consciousness and paralysis. Oral or injected calcium solution is a common treatment.

Molt The shedding and renewal of a chicken's feathers on a roughly annual basis. Egg production slows or stops during the molt.

Must The juice of crushed grapes or other fruit before fermentation.

Nest egg A decoy egg of plastic or wood used to attract layers to nesting boxes.

Newcastle disease A contagious respiratory disease common in poultry but preventable by vaccination.

pH A measure of acidity or alkalinity. Values range from 0 to 14. A food is neutral when its pH is 7.0. Lower values are increasingly more acidic; higher values are increasingly more alkaline.

Parkway The area between the street curbing and the sidewalk.

Pecking order The social hierarchy established by chickens through pecking.

Pectin A natural substance available in powder or liquid form that jells fruit for jams and jellies.

Perennial A plant that lives for three or more years and generally flowers each year. By perennial, gardeners usually mean herbaceous perennial, although woody plants such as vines, shrubs, and trees are also perennial.

Picking The instinctual trait among chickens to establish hierarchy by pecking. Also, attacking any sign of blood on another chicken.

Pickling Adding enough vinegar or lemon juice to a low-acid food to lower its pH to 4.6 or less. Properly pickled foods may be safely heat processed in boiling water. Salt can also be used for pickling.

Pickout Damage done to a chicken's vent by the pecking of other chickens.

Pollen bound A brood nest so filled with pollen that the queen has no room for laying eggs.

Pressure canner A specifically designed kettle with a lockable lid used for heat processing low-acid food under pressure. Such canners have jar racks, one or more safety devices, systems for exhausting air, and a way to measure and control pressure.

Propolis Sometimes called "bee glue", bees produce this sticky sealant from tree bark and foliage and use it to mend cracks in the hive.

Ruminant An animal with a multi-chambered stomach that regurgitates and rechews food to facilitate digestion.

Scours Severe diarrhea in livestock, a condition sometimes fatal to young animals.

Spent hen A laying hen that has reached the end of her productive life.

Straight run Newly hatched chicks that have not been separated by sex.

Stripping A final step in milking to remove the last vestiges of milk in the teat.

Queen The sole egg-laying member of a bee hive, the queen is created by feeding royal jelly to a larva. A hive has but one reigning queen. She lives 2 to 3 years.

Queen cage A miniature shipping crate for a queen bee and a small contingent of workers.

Queen cage candy The plug on a queen cage, used as food in shipment and a slow-release exit once the cage is placed in the hive.

Queen excluder A barrier made of wood, wire, or metal that confines the queen, but not workers, to the brood nest.

Plat survey A detailed drawing of a lot produced by a surveyor, with exact locations of lot lines, buildings, walkways, and driveways. A plat survey accompanies title documents.

Range-fed Chickens that graze on pasture.

Raw packed See *Cold packed.*

Royal jelly A highly nutritious milky secretion produced by nurse bees and fed to the queen and young larvae. Feeding larvae exceptional amounts of Royal jelly causes them to morph into queens.

Site plan A simple drawing that includes all lot lines, buildings, walkways, and fences on a property.

Succession planting A technique in which an early, fast-maturing crop, such as lettuce or peas, is followed by another crop during the same growing season. Boosts productivity per square foot of growing space.

Supers The boxes that hold frames, added to a hive to provide more room for honey storage.

Top-fermenting yeast Effective at warm temperatures and able to tolerate high alcohol concentration, top-fermenting yeast yields sweeter beer. Also known as "ale yeast."

Underground feeder (UF) cable Outdoor cable approved for direct burial.

Unthrifty Any animal that for genetic or health reasons does not grow at the normal rate.

Wether A castrated male goat.

Worker bee An adult female bee that gathers pollen and nectar and builds combs to produce honey. A worker may live for only a few weeks in the summer but several months if born in winter.

Wort The liquid made of malted barley that is the first stage in beer making. It contains the sugars on which yeast works to produce alcohol.

Yearling A young goat between 6 and 12 months of age.

Zoning Municipal regulations concerning changes that may infringe on a neighbor or could affect neighborhood property values.

Zymurgy The study of the biochemical process involved in brewing and fermentation.

Index

Index

Index

Index

Credits

page 2: Chris Price/iStockphoto **page 6:** *left* cpaquin/iStockphoto; *center* Jiri Vaclavek/Dreamstime; *right* luna4/iStockphoto **page 7:** *left and right* Greenleaf Publishing, Inc.; *center* Mike Rodriguez/iStockphoto **page 8:** *left* Lucian Coman/Dreamstime; *right* Ben Toht, Greenleaf Publishing, Inc. **page 9:** Greenleaf Publishing, Inc. **page 10:** *top* Ben Toht, Greenleaf Publishing, Inc.; *bottom left* Maran/Dreamstime; *bottom center* urbancow/iStockphoto; *bottom right* Elina Manninen/Dreamstime **page 11:** *left* Chris Price/iStockphoto; *top right* Valarie Staus/iStockphoto; *bottom right* Greenleaf Publishing, Inc. **page 14:** Greenleaf Publishing, Inc. **page 15:** *center* René Mansi/iStockphoto; *bottom right* Greenleaf Publishing, Inc. **page 16:** *center right* cpaquim/iStockphoto **page 17:** *top* kkgas/iStockphoto; *center* Greenleaf Publishing, Inc.; *bottom* Chris Price/iStockphoto **pages 18–19, 21–22:** Greenleaf Publishing, Inc. **page 23:** *top* Eclypse78/Dreamstime; *bottom* Greenleaf Publishing, Inc. **page 24:** *top* Paul Senyszyn/iStockphoto; *bottom* Chris Price/iStockphoto **page 25:** Alice Joyce **page 26:** Greenleaf Publishing, Inc. **page 27:** *top* Toro; *bottom* Greenleaf Publishing, Inc. **pages 28–29:** Greenleaf Publishing, Inc. **page 34:** Chris Price/iStockphoto **page 35** *top* Melanie Defazio/Dreamstime; *bottom* Greenleaf Publishing, Inc. **pages 40, 42–43:** Greenleaf Publishing, Inc. **page 44:** Christine E. Wigand **page 45:** *top* Kkgas/iStockphoto; *bottom* Nicky Gordon/iStockphoto **page 48:** *top* Liza McCorkle/iStockphoto; *center* Patrick Laverdant/iStockphoto; *bottom* Patricia Nelson/iStockphoto **page 49:** Greenleaf Publishing, Inc. **page 50:** Chris Price/iStockphoto **page 51:** *top left* Marpalusz/Dreamstime; *top center* Yykkaa/Dreamstime; *top right* Dinamir Predov/iStockphoto; *center right* Alison Stieglitz/iStockphoto **page 52:** Elena Moiseeva/Dreamstime **page 53:** *top left* Yobiduba/Dreamstime; *top center* Joan Kimball/iStockphoto; *top right* Chuyu/Dreamstime **page 54:** *top left* Igor Golovniov/Dreamstime; *bottom left* Allihays/Dreamstime; *bottom mid left* Robert Keenan/Dreamstime; *bottom mid right* Leszek Wilk/Dreamstime; *bottom right* Norm Stangl/Dreamstime **page 56:** Margojh/Dreamstime **page 57:** *top left* Lin Jyun Zuo/Dreamstime; *top center* Onionhead/Dreamstime; *top right* Darko Plohl/Dreamstime; *center* Stepan Popov/Dreamstime; *center right* Amy Nicolai/Dreamstime; *bottom right* Leva Zemite/Dreamstime **page 58:** *top left* Marianne Lachance/Dreamstime; *top mid left* Godrick/Dreamstime; *top mid right* Bncc 369/Dreamstime; *top right* Constantin/Dreamstime **page 59:** *top* Greenleaf Publishing, Inc.; *bottom* Marekp/Dreamstime **page 60:** *top* Bluestock/Dreamstime; *bottom* Darko Plohl/Dreamstime **page 61:** *top* Dmitry Mizintsev/Dreamstime; *bottom* Gymane/Dreamstime **page 62:** *left* Omar Mashaka/Dreamstime; *right* Norman Chan/Dreamstime **page 63:** *left* Eagle/Dreamstime; *right* Inacio Pires/Dreamstime **page 62:** *top* Greenleaf Publishing, Inc.; *left center* Jose Tejo/Dreamstime; *right center* LLC54613/Dreamstime **page 65:** Ieva Zemite/Dreamstime **page 72:** *left* Lilli Day/iStockphoto; *right* Kelly Cline/iStockphoto **page 73:** Ursula Alter/iStockphoto **page 74:** *top* Damithri/Dreamstime; *bottom* dirkr/iStockphoto **page 75:** *top* Lianem/Dreamstime; *bottom* Nadia

Vlashchenko/Dreamstime **page 76:** *top* 3desc/Dreamstime; *bottom* Coldfusion/Dreamstime **page 77:** *top* Alie Van Der Velde-Baron/Dreamstime; *bottom* Elkeflorida/Dreamstime **page 78:** *top* Hmproudlove/Dreamstime; *bottom* Antiode/Dreamstime **page 79:** *top* Darko Plohl/Dreamstime; *bottom* Stacey Lynn Payne/Dreamstime **page 80:** *top* Babar760 Dreamstime; *center* Donnarae/Dreamstime; *bottom* Elfthryth/Dreamstime **page 82:** *top* ruhrpix/iStockphoto; *center* Jasuaol/Dreamstime; *bottom* Marshall Turner/Dreamstime **page 84:** Jiri Vaclavek/Dreamstime **page 85:** Fotomy/Dreamstime **page 86:** Joop Snijder/Dreamstime **page 87:** Dennis Richardson/Dreamstime **page 88:** Darko Plohl/Dreamstime **page 92:** *left* Darko Plohl/Dreamstime; *center* Franz Schlögl/Dreamstime **page 93:** *left* Xiye/Dreamstime; *right* Martin Green/Dreamstime **page 94:** *top* Anikasalsera/Dreamstime; *center* John Kershner/Dreamstime; *bottom* Greenleaf Publishing, Inc. **page 95:** Blacksnake/Dreamstime **page 96:** Elina Manninem/Dreamstime **page 97:** *top* Jirkal3/Dreamstime; *center* Irina Kodentseva Dreamstime **page 98:** *top* Adela Manea/Dreamstime; *bottom* Joselito Briones/iStockphoto **page 99:** Svetlana Larina/Dreamstime **page 100:** *top* Lianem/Dreamstime; *bottom* Tatyana Shkondina/Dreamstime **page 101:** Michael Jost/Dreamstime **page 105:** *top* Judith Bicking/Dreamstime; *bottom* Jiri Vaclavek/Dreamstime **page 106:** *top* Twila Madison/Dreamstime; *bottom* Sprokop/Dreamstime **page 107:** *left* Greenleaf Publishing, Inc.; *right* Lindsey Davis/Dreamstime **page 108:** Greenleaf Publishing, Inc. **page 111:** *top* Kenneth D. Duiden/Dreamstime; *bottom* Carolecastelli/Dreamstime **page 112:** *top* Susan Legget/Dreamstime; *bottom left* Ben Goode/Dreamstime; *bottom right* Phillip Minnis/Dreamstime **page 113:** *top* Gordana Sermek/Dreamstime; *bottom* Olga Miltsova/Dreamstime **page 116:** *top* Igorr/Dreamstime; *bottom* Kathy Puckett/iStockphoto **page 117:** Greenleaf Publishing, Inc. **page 118:** *top* Victorita/Dreamstime; *center* Juan Moyano/Dreamstime; *bottom* Fausto Fiori/Dreamstime **page 119:** *top* Yosef Erpert/Dreamstime; *center* Teresa Kearney/Dreamstime; *bottom* Darko Plohl/Dreamstime **pages 120–121:** Greenleaf Publishing, Inc. **page: 122:** Ben Toht Greenleaf Publishing, Inc. **page 123:** Greenleaf Publishing, Inc. **page 124:** David Claassen/iStockphoto **page 124:** *top left, top right, center left* Greenleaf Publishing, Inc.; *center right* Ben Toht Greenleaf Publishing, Inc.; *bottom left* Christopher Moncrieff/Dreamstime; *bottom right* Marilyn Barbone/Dreamstime **page 128:** Ben Toht Greenleaf Publishing, Inc. **page 129:** David Bissette, CatawbaCoops.com **page 130:** *top* Shawn Smith; *bottom* Greenleaf Publishing, Inc. **page 131:** *top* Chicken Coop Source; *bottom* Greenleaf Publishing, Inc. **page 132:** *top* johnnyscriv/iStockphoto; *center* Renee Fraser; *bottom* Omlet, Ltd. **page 133:** *top* Greenleaf Publishing, Inc.; *bottom* Hayneedle.com **page 134:** Steve Orth, Eden Creek Farm **pages 135–137:** Greenleaf Publishing, Inc. **page 138:** Ben Toht Greenleaf Publishing, Inc. **page 139:** Greenleaf Publishing, Inc. *except bottom right* Ben Toht **page 140:** Ben Toht Greenleaf Publishing, Inc. **pages 141–144:** Greenleaf Publishing, Inc. **page 145:** fotokon/iStockphoto **page 146:** *left* Nancy Tripp/Dreamstime; *right* Duncan

Noakes/Dreamstime **page 147:** *left* Greenleaf Publishing, Inc.; *right* Kimberpix/Dreamstime **page 148:** *left* Igor Stevanovic/Dreamstime; *right* Krodere/iStockphoto **page 149:** Lana Langlois/Dreamstime **pages 150–151:** Greenleaf Publishing, Inc. **page 152:** *top* Greenleaf Publishing, Inc.; *bottom* National Pygmy Goat Association **page 154:** Greenleaf Publishing, Inc. **page 155:** *top left and right* Eagle Sheds; *bottom* Patti Christmas **page 156:** *top left* Static 85/Dreamstime; *top right, bottom left* Greenleaf Publishing, Inc.; *bottom right* Tom Curtis/Dreamstime **page 157:** National Pygmy Goat Association **page 158:** *top left* David Watts Jr./Dreamstime; *top right* Joshua and Gurney Davis; *bottom left* Ravens Rest Farm; *bottom right* 578foot/iStockphoto **page 159:** *top* National Pygmy Goat Association; *bottom* Greenleaf Publishing, Inc. **pages 160, 161:** Greenleaf Publishing, Inc. **page 162:** *top* Greenleaf Publishing, Inc.; *bottom* National Pygmy Goat Association **page 163:** Jennifer Mackenzie/iStockphoto **pages 164, 166–167:** Greenleaf Publishing, Inc. **page 168:** *top* Southside Farms; *bottom* American Dexter Cattle Association **page 169:** *left* Tupungato/Dreamstime; *right* Will Iredale/Dreamstime **page 170:** Mike Rodriguez/iStockphoto **page 171:** Ben Toht Greenleaf Publishing, Inc. **page 172:** *top* American Bee Federation; *bottom* Stefanie Booth **page 173:** Stefanie Booth **page 174:** *top* Greenleaf Publishing, Inc.; *bottom left* Ben Toht; *bottom right* Karen Massier/iStockphoto **page 176:** *top* Schuyler Smith; *bottom* Bill Rawleigh, The Garden Hive **page 178:** Milkwood.net **page 179:** Ernie Schmidt **pages 180–181:** Greenleaf Publishing, Inc. **pages 182–185:** Stefanie Booth **page 186:** Milkwood.net **page 187:** *top* Milkwood.net; *bottom* Ben Toht Greenleaf Publishing, Inc. **page 188:** Karen E. Bean, Brookfield Farm **page 189:** Paul Tucker **page 190:** Greenleaf Publishing, Inc. **page 191:** Mikko Pitkänen/Dreamstime **page 193:** Greenleaf Publishing, Inc. **page 194:** *top* YinYang/iStockphoto; *bottom* Greenleaf Publishing, Inc.; **page 195:** National Presto Industries, Inc. **pages 196–199:** Greenleaf Publishing, Inc. **page 200:** *top* National Presto Industries, Inc.; *bottom* Liudmila Sundikova/Dreamstime **page 201:** Greenleaf Publishing, Inc. **page 202:** Elena Elisseeva/Dreamstime **page 203:** Buccaneer/Dreamstime **page 204:** Colorvsbw/Dreamstime **page 205:** Lorna/Dreamstime **page 206:** Greenleaf Publishing, Inc. **page 207:** Robyn Mackenzie/Dreamstime **page 208:** Alexander Podshivalov/Dreamstime **page 209:** Olesia Sarycheva/Dreamstime **page 210:** Boddau Wankowicz/Dreamstime **211:** Hedda Gjerpen/iStockphoto **page 212:** Paul Prescott/iStockphoto **page 213:** *top* Excalibur Products; *bottom* Greenleaf Publishing, Inc. **pages 214, 217:** Excalibur Products **page 218:** Leena Damie/Dreamstime **page 221:** Greenleaf Publishing, Inc. **page 226:** Daseaford/Dreamstime **page 229:** Jon Larson/iStockphoto **page 230:** Joel Sommer, WinePress.US **page 231:** Bert Folsom/Dreamstime **page 232:** Valarie Staus/iStockphoto **page 233:** Parker Dean/iStockphoto **page 234:** Tom Eagan/iStockphoto **page 339:** YinYang/iStockphoto **page 241:** Olga Lyubkin/Dreamstime **Back cover:** *in descending order* cpaquin/iStockphoto; luna/iStockphoto; Greenleaf Publishing, Inc.; Mike Rodriguez/iStockphotov

Have a gardening or landscaping project?
Look for these and other fine Creative Homeowner books
wherever books are sold

ULTIMATE GUIDE TO OUTDOOR PROJECTS
Hardscape and landscape projects that add value and enjoyment to your home.

Over 1,200 photographs and illustrations.
368 pp.
8½" × 10⅞"
$19.95 (US)
$23.95 (CAN)
BOOK #: CH277873

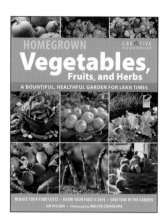

HOMEGROWN VEGETABLES
A complete guide to growing your own vegetables, fruits, and herbs.

Over 275 photographs and illustrations.
192 pp.
8½" × 10⅞"
$16.95 (US)
$20.95 (CAN)
BOOK #: CH274551

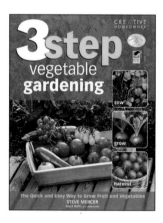

3 STEP VEGETABLE GARDENING
A quick and easy guide for growing your own fruit and vegetables.

Over 300 photographs.
224 pp.
8½" × 10⅞"
$19.95 (US)
$21.95 (CAN)
BOOK #: CH274557

SUCCESSFUL CONTAINER GARDENING
Information to grow your own flower, fruit, and vegetable "gardens."

Over 240 photographs.
160 pp.
8½" × 10⅞"
$14.95 (US)
$17.95 (CAN)
BOOK #: CH274857

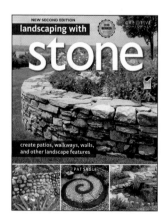

LANDSCAPING WITH STONE
Ideas for incorporating stone into the landscape.

Over 335 photographs.
224 pp.
8½" × 10⅞"
$19.95 (US)
$21.95 (CAN)
BOOK #: CH274179

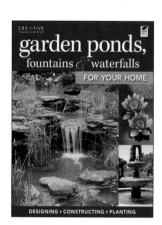

GARDEN PONDS, FOUNTAINS & WATERFALLS FOR YOUR HOME
Secrets to creating garden water features.

Over 490 photographs and illustrations.
256 pp.
8½" × 10⅞"
$19.95 (US)
$22.95 (CAN)
BOOK #: CH274450

For more information and to order direct, go to **www.creativehomeowner.com**